AF333807

CELL BIOLOGY RESEARCH PROGRESS

KERATINOCYTES

STRUCTURE, MOLECULAR MECHANISMS AND ROLE IN IMMUNITY

CELL BIOLOGY RESEARCH PROGRESS

CELL BIOLOGY RESEARCH PROGRESS

KERATINOCYTES

STRUCTURE, MOLECULAR MECHANISMS AND ROLE IN IMMUNITY

ELIA RANZATO
EDITOR

New York

For permission to use material from this book please contact us:
Telephone 631-231-7269; Fax 631-231-8175
Web Site: http://www.novapublishers.com

Library of Congress Cataloging-in-Publication Data

ISBN: 978-1-62618-798-6

Library of Congress Control Number: 2013938560

Published by Nova Science Publishers, Inc. † New York

Contents

Preface

It is my very great pleasure to introduce the book "Keratinocytes: Structure, Molecular Mechanisms and Role in Immunity."

As we know, the skin is the largest organ of the body, and above all its multiple tasks, it provides a protective barrier between internal and external environments. The various functions of skin are reflected by its complex structure and composition. The skin is composed of epidermis and underlying dermis. The epidermis is composed of the outermost layers of the skin, and keratinocytes are the major cells.

The ability of keratinocyte to proliferate, differentiate and migrate plays a pivotal role for normal regeneration of the skin, but also during injury and inflammation processes. The repair process of the skin is normally subdivided, as reported by *Simona Martinotti* and *Elia Ranzato* in *"Dynamic interplay between cell types during wound healing"*, into a series of events involving tightly regulated cellular and molecular events.

Tomohisa Hirobe in *"Regulation of the Proliferation and Differentiation of Keratinocytes and Cellular Interaction between Keratinocytes and the Tissue Environment"* describes the most important molecules in keratinocyte differentiation, released in autocrine fashion, but also as paracrine regulation by factors derived from melanocytes and fibroblasts. Keratinocytes also produce and release proliferation and differentiation-stimulating factors able to modify the phenotype of melanocytes and fibroblasts. This double paracrine regulation seems to play an important role in the maintenance of skin homeostasis.

Tanja Popp and *Christian Ries* in *"Keratinocyte differentiation: Focus on signal transduction"* provide insight into the molecular mechanisms governing the differentiation of interfollicular stem cells into mature keratinocytes. This differentiation process is controlled by a calcium (Ca^{2+}) gradient and characterized by the expression of specific marker proteins. The MAPK signaling cascades are the principal players in the control of proliferation and differentiation in immature keratinocytes. Dysregulation of this complex differentiation program can contribute do several pathological conditions, such as psoriasis.

In the skin, the interaction between keratinocytes and fibroblasts is important for normal development and regeneration after wounding, and epithelial-mesenchymal interactions are important for integrity of skin. *Ekaterina A Vorotelyak* and coworkers describe in "*Epithelial-to-mesenchymal transition in epidermal keratinocytes*" the cellular and molecular details of this process that allow achieving a tremendous keratinocyte plasticity, which depends on the cross talk of intercellular signaling pathways and environmental cues.

The skin is also the primary barrier between host and environment, and a huge number of microbes colonize on skin and contact with keratinocyte. However, the skin normally is not infected or inflamed. So, keratinocytes as describes by *Yuping Lai* in "*Keratinocytes: gatekeeper in innate defense*" participate actively in cutaneous immune responses to protect the skin from infections, functioning as a gatekeeper.

In addition to providing a physical barrier, keratinocytes represent an immunological barrier to prevent the invasion of pathogen. As described by *Juan L. Rendon* and *Katherine A. Radek* in "*Mechanisms for keratinocyte immune defense: implications for cutaneous wound healing*" keratinocytes utilize various pathogen recognition receptors, including toll-like receptors, to distinguish pathogens from host molecules. Some pathologic skin conditions, such as rosacea and psoriasis, are characterized by defects in skin immunity and subsequent impaired barrier function, so making keratinocytes potential therapeutic targets for these conditions.

Giovina Vianale and coworkers in "*Keratinocytes and anti-microbial peptides: influence of Extremely Low Frequency Electromagnetic Field (ELF-EMF)*" report about the secretion of antimicrobial peptides, a conserved innate host-defense mechanism, by keratinocytes and how the exposure to low-frequency electromagnetic fields affects the cytokines and anti-microbial peptides expression, and how this could speed up and improve tissue repair.

In: Keratinocytes
Editor: Elia Ranzato

ISBN: 978-1-62618-798-6
© 2013 Nova Science Publishers, Inc.

Chapter 1

Dynamic Interplay between Cell Types during Wound Healing

Simona Martinotti and Elia Ranzato[*]
DiSIT - Dipartimento di Scienze e Innovazione Tecnologica,
Università del Piemonte Orientale "Amedeo Avogadro", Italy

Abstract

Wound healing is a complex biological process requiring regulated interactions between cells, local and systemic growth factors and the extracellular matrix components in which these entities interact.

Tissue restoration involves several biochemical processes of tissue repair like granular tissue formation, matrix deposition, and re-epithelialization. All these events involve active participation of various cell types and their cross talk plays an important role during the wound healing process.

We summarize the interplays between keratinocytes and others cells, such as endothelial cells, fibroblasts and inflammatory cells, and how these cell types impact to control the re-epithelialization, angiogenesis and the extracellular matrix remodeling. The best knowledge of this process will open new therapeutic interventions on skin wounds.

---

[*] E-mail: ranzato@unipmn.it.

Introduction

Despite significant progress that has been achieved in our understanding of normal wound healing process and the correlated pathologies that lead to wound chronicity, chronic wounds of different etiology remain a significant health care burden. Wound healing is a complex process that can be roughly divided into 3 overlapping phases: inflammatory, proliferation and remodeling (Singer and Clark, 1999). Inflammation is a highly effective component of the body reaction to the injury. This process normally leads to tissue repair and restoration of function (Eming et al., 2007). Bleeding occurs immediately after tissue injuries as a result of the disruption of blood vessels. The first part of wound healing is the hemostasis that consists of two major processes: development of a fibrin clot and coagulation. Platelets are first cells to appear into wound bed, and exposure to the extracellular matrix in the vascular wall induces platelet activation. Upon activation, platelets undergo adhesion as well as aggregation and at the same time release some mediators (e.g. serotonin, thromboxane A_2, etc.) and adhesive proteins (such as fibrinogen, fibronectin, etc.). With the conversion by thrombin of fibrinogen to fibrins, a fibrin clot is formed to stop the bleeding. The second component of hemostasis is coagulation, achieved by intrinsic and extrinsic coagulation pathway, leading to transformation of prothrombin into thrombin that in turn converts soluble fibrinogens to insoluble fibrin. Damaged tissue releases a lipoprotein known as tissue factor, which activates the extrinsic coagulation pathway (Monaco and Lawrence, 2003). Platelets do not perform only critical functions of hemostasis, but contribute significantly to other processes of wound healing, including inflammation, re-epithelialization, fibroplasia and angiogenesis (Ranzato et al., 2010). Platelets influence wounds through an infiltration of leukocytes by releasing platelet-derived chemotactic factors, and promote tissue regeneration by releasing several growth factors that play a pivotal role in wound repair (Ranzato et al., 2010).

Once bleeding is controlled, inflammatory cells migrate into wound (chemotaxis) and promote the inflammatory phase, which is characterized by the sequential infiltration of neutrophils, macrophages, and lymphocytes (Campos et al., 2008). Immediately after injury, neutrophils and monocytes migrate from capillaries into the wounded tissue, with neutrophils being the first to arrive in great numbers. Later in inflammation, the number of neutrophils declines and macrophages (tissue-derived monocytes) predominate. Neutrophils and monocytes are recruited by chemotactic factors released during hemostasis and by mast cells. Some growth factors released by platelets, such as PDGF and TGF-b, are also potent chemotactic factors for leukocytes. A critical function of neutrophils is the clearance of invading microbes and cellular debris in the wound area, although these cells also produce substances such as proteases and reactive oxygen species (ROS), which cause some additional bystander damage. Macrophages are considered to be the most important regulatory cell in the inflammatory reaction. After binding to the extracellular membrane, bacterial, cellular, and tissue phagocytosis, and subsequent destruction, are accomplished through the release of biologically active oxygen intermediates and enzymatic proteins.

Macrophages release chemotactic factors (eg, fibronectin) that attract fibroblasts to the wound area. New blood vessel growth follows a gradient of angiogenic factors produced by hypoxic macrophages because macrophages do not produce these angiogenic factors when either fully oxygenated or anoxic. Macrophages can be considered as factories for growth factor production, including PDGF, fibroblast growth factor, vascular endothelial growth factor, TGF-b, and TGF-a (Ranzato and Burlando, 2011). As macrophages clearthese apoptotic cells, they undergo a phenotypic transition to a reparative state that stimulates keratinocytes, fibroblasts, and angiogenesis to promote tissue regeneration (Meszaros et al., 2000). In this way, macrophages play a pivotal role in the transition between inflammation and repair phase. Several studies have analyzed skin wound healing upon macrophage depletion in order to understand the implication of macrophages in the process (Rodero and Khosrotehrani, 2010). In a model of null mice where animals lacked macrophages, mast cells and functional neutrophils due to defective myelopoiesis, wounds performed on newborns additionally treated with antibiotic healed at the same speed as wild types, but without scar formation, suggesting that inflammatory cells are not needed for wound closure (Martin et al., 2003). However, several recent models of specific inducible macrophage depletionresulted in detrimental effect of pre-injury depletion of macrophages (Di Pietro et al., 2003;Lucas et al., 2010). Mice depleted before injury typically show a defect in re-epithelialization, granulation tissue formation, angiogenesis, wound cytokine production and myofibroblast associated wound contraction. The appearance of macrophages is followed somewhat later by T-lymphocytes, which peak numbers occurduring the late-proliferative/early-remodeling phase. The role of T-lymphocytes is not completelyunderstood and is a current area of investigation. Interestingly, recent studies in mice deficient in both T- and B-cells have shown that scar formation is diminished in the absence oflymphocytes (Gawronska-Kozak et al., 2006). Skin gamma-delta T-cells, also known as dendritic epidermal T-cells (DETC), arising from fetal thymic precursors, regulate many aspects of wound healing, including maintaining tissue integrity, defending against pathogens, and regulating inflammation. DETC are activated by stressed, damaged, or transformed keratinocytes and produce fibroblast growth factor 7 (FGF-7), keratinocyte growth factors, and insulin-like growth factor-1, to support keratinocyte proliferation and cell survival (Macleod and Havran, 2011). The location of DETC in barrier tissues, such as the skin, lung, and intestine, makes them an ideal candidate for participating in protecting the organism from infection and maintaining tissue homeostasis (Havran and Jameson, 2010). Some studies have demonstrated that DETC produce chemokines and cytokines that contribute to the initiation and regulationof the inflammatory response during wound healing. Little is known about the importance of cross-talk between DETC and keratinocytes to the maintenance of normal skin and wound healing. Mice lacking or defective in skin gamma-delta T-cells show a delay in wound closure and a decrease in the proliferation of keratinocytesat the wound site (Jameson and Havran, 2007; Mills et al., 2008).

Following completion of their tasks, neutrophils and monocytes must be eliminated in order to initiate the next stage of wound healing. Rapid increase in cell infiltration during tissue reconstruction is balanced by apoptosis. The specific mechanisms of activation of apoptotic process and the regulation from the wound site remain poorly understood (Desmouliere et al., 1995). Phagocytosis of apoptotic cells has distinctive morphologic features and unique downstream consequences. Some authors coined the term efferocytosis (deCathelineau and Henson, 2003; Gardai et al., 2006) in respect to phagocytosis of apoptotic cells. Inappropriate clearance of cell corpses may lead to autoimmune diseases and chronic inflammation (Rosen and Casciola-Rosen, 1999). Most of the symptoms associated with inflammatory response last for approximately 2 weeks. If inflammation persists for months or years, it is called chronic inflammation. Chronic inflammation often occurs when a wound is sealed by necrotic tissue, is contaminated with pathogens, or contains foreign materials that cannot be phagocytized or solubilized during the acute inflammatory phase. The body responds to the presence of persistent foreign material and/or infection by local proliferation of mononuclear cells. Non-resolving persistent inflammation may derail the healing cascade resulting in chronic wounds (Meszaros et al., 2000). Macrophages that have ingested foreign material will remain in the tissue if they are unable to solubilize the ingested material. Macrophages attract fibroblasts and over time may produce increased quantities of collagen, leading to a slowly forming encapsulated mass of fibrous tissue, a granuloma (Khanna et al., 2010). The initial inflammatory response provides the necessary framework to the production of a new functional barrier, and in second phase the cellular activity predominates. The major events during this phase are the creation of a permeability barrier (i.e., re-epithelialization), the establishment of appropriate blood supply (ie, angiogenesis), and reinforcement of the injured dermal tissue (i.e., fibroplasia) (Martinotti, 2012). An essential feature of a healed wound is the restoration of an intact epidermal barrier through wound epithelialization, also known as re-epithelialization. A complex balance of signaling factors and surface proteins are expressed and regulated in a very tightly regulated manner in order to orchestrate wound re-epithelialization. Re-epithelialization of the wound is the result of three overlapping processes: migration, proliferation, and differentiation of keratinocytes. Keratinocyte migration is an early event in wound re-epithelialization (Raja et al., 2007). Approximately 12 hours after wounding, keratinocytes are activated with dissolution of cell-cell and cell-substratum contacts.

The keratinocyte proliferation rate during migration is inhibited. Moreover, keratinocytes actively migrating may show phenotypic shift in the repertoire of expressed integrins, such as CD44 and some markers usually expressed by squamous cells, as well as keratinocyte express a different array of keratins than do those residing in intact epidermis (Larjava et al., 1993; Stoler et al., 1989). Growth factors and cytokines regulate re-epithelialization process, orchestrating a complex and dynamic balance with the integration of signals in the wound healing environment (Werner and Grose, 2003). Another important element in the re-epithelialization process is the ability of keratinocytes to detach from the underlying basal lamina and migrate through the fibrin and provisional matrix present in wound bed. Migrating keratinocytes produce MMPs, such as MMP-9, which specifically degrades type IV collagen and laminins in the basement membrane (Parks, 1999). The important role of MMPs is underlined by experiments that have shown as exogenous treatment with MMP inhibitors delay wound healing and re-epithelialization process (Pilcher et al., 1999). By contrast, overexpression of the MMPs may delay the wound healing process and have been shown to be elevated in chronic wounds (Xue et al., 2006).

When migration stops, as a result of contactinhibition, keratinocytes reattach themselves to the underlying matrix, reconstitute the basement membrane, and then resume the process of terminal differentiation to generate a stratified epidermis (Laplante et al., 2001). Dermal reconstitution begins approximately 3 to 4 days after injury, which includes new blood vessel formation, and the accumulation of fibroblasts and matrix, named fibroplasia. Fibroplasia describes a process of fibroblast proliferation, migration into wound fibrin clot, and production of new collagen and other matrix proteins, such as glycosaminoglycans and proteoglycans (Nirodi et al., 2000). After the migration of fibroblasts into the wound, they gradually change to profibrotic phenotypes. Fibroblasts are also modulated into phenotypes of myofibroblastsand participate in wound contraction (Clark, 1993). Successful wound healing depends upon angiogenesis, the growth of new capillary blood vessels. Clinically, new capillaries first become visible in the wound bed 3–5 days after injury (Tonnesen et al., 2000). Angiogenesis involves a phenotypic alteration of endothelial cells, directed migration, and various mitogenic stimuli. Cytokines released by macrophages stimulate angiogenesis during wound healing, as well as low-oxygentension, lactic acid, and biogenic amines produced in the wound (Remensnyder and Majno, 1968). Growth factors are critical mediators of wound neovascularization expressed during healing in a temporal, orchestrated fashion. By initiating complex signaling pathways in vascular endothelial cells, growth factors activate cells to undergo proliferation, migration, tube formation, and vascular maturation. PDGF-b and VEGF are among the most critical known angiogenic growth factors, because they induce blood vessel growth and guide vascular maturation (Nissen et al., 1998).

Some investigations have shown that the wound extracellular matrix (ECM) can regulate angiogenesis in part by modulating integrin receptor expression (Serini et al., 2006). Understanding the molecular mechanisms that regulate wound angiogenesis, particularly how ECM modulates ECM receptor and angiogenic factor requirements, may provide new approaches for treating chronic wounds. Following robust proliferation and ECM synthesis, wound healing enters the final remodeling phase, which can last for years. One critical feature of the remodeling phase is ECM remodeling to an architecture approaching that of the normal tissue. The wound also undergoes physical contraction throughout the entire wound healing process, which is believed to be mediated by contractilefibroblasts (myofibroblasts) that appear in the wound (Gosain and DiPietro, 2004). One of the characteristics of wound remodeling is the change of ECM composition. Collagen fibers constitute approximately 80% of the dry weight of normal human dermis (Leung et al., 2012), while type I collagen accounts for approximately 80% and type III collagen is only 10% of total collagens. During wound healing, however, type III collagen is the predominant collagen synthesized by fibroblasts (Ranzato et al., 2011). With wound closure, a gradual turnover of collage occurs, and type

III collagen is degraded and type I collagen synthesis increases (Booth et al., 1980).

The regulation of wound contraction remains poorly defined. Information regarding the effects of specific cytokines on contraction is limited and often conflicting. TGF-b has been found to promote contraction even in the absence of serum (Finesmith et al., 1990); PDGF-b has also been found to either increase contraction (Clark et al., 1989) or have no effect (Montesano and Orci, 1988), while both FGF and EGF have been found by different authors to either have no effect or cause a moderate enhancement of contraction (Montesano and Orci, 1988). The process of wound healing is essentially similar in all tissues and is relatively independent of the mode of injury; however, slight variations in the relative contribution of the different elements to the overall result may occur. During the healing process numerous cytokines and growth factors are produced by the various cell types present in the wound area. The level of production of the different cytokines depends on the regulation of the cross talk between the major cell populations in the wound: epithelial cells, endothelial cells, fibroblasts and inflammatory cells (Ranzato and Burlando, 2011). However, the literature lacks *in vivo* characterization of the cytokine production kinetics by the various cell types during skin wound healing and most of our knowledge concerning the ability of a cell population to produce or to respond to a specific cytokine is based on *in vitro* studies.

References

Booth BA, Polak KL, Uitto J. 1980. Collagen biosynthesis by human skin fibroblasts. I. Optimization of the culture conditions for synthesis of type I and type III procollagens. *Biochimicaetbiophysicaacta* 607:145-60.

Campos AC, Groth AK, Branco AB. 2008. Assessment and nutritional aspects of wound healing. Current opinion in clinical nutrition and metabolic care 11:281-8.

Clark RA. 1993. Basics of cutaneous wound repair. *The Journal of dermatologic surgery and oncology* 19:693-706.

Clark RA, Folkvord JM, Hart CE, Murray MJ, McPherson JM. 1989. Platelet isoforms of platelet-derived growth factor stimulate fibroblasts to contract collagen matrices. *The Journal of clinical investigation* 84:1036-40.

deCathelineau AM, Henson PM. 2003. The final step in programmed cell death: phagocytes carry apoptotic cells to the grave. *Essays in biochemistry* 39:105-17.

Desmouliere A, Redard M, Darby I, Gabbiani G. 1995. Apoptosis mediates the decrease in cellularity during the transition between granulation tissue and scar. *The American journal of pathology* 146:56-66.

Di Pietro SM, Corsico B, Perduca M, Monaco HL, Santome JA. 2003. Structural and biochemical characterization of toad liver fatty acid-binding protein. *Biochemistry* 42:8192-203.

Eming SA, Krieg T, Davidson JM. 2007. Inflammation in wound repair: molecular and cellular mechanisms. *The Journal of investigative dermatology* 127:514-25.

Finesmith TH, Broadley KN, Davidson JM. 1990. Fibroblasts from wounds of different stages of repair vary in their ability to contract a collagen gel in response to growth factors. *Journal of cellular physiology* 144:99-107.

Gardai SJ, Bratton DL, Ogden CA, Henson PM. 2006. Recognition ligands on apoptotic cells: a perspective. *Journal of leukocyte biology* 79:896-903.

Gawronska-Kozak B, Bogacki M, Rim JS, Monroe WT, Manuel JA. 2006. Scarless skin repair in immunodeficient mice. Wound repair and regeneration : *official publication of the Wound Healing Society -and] the European Tissue Repair Society* 14:265-76.

Gosain A, DiPietro LA. 2004. Aging and wound healing. *World journal of surgery* 28:321-6.

Havran WL, Jameson JM. 2010. Epidermal T cells and wound healing. *Journal of immunology* 184:5423-8.

Jameson J, Havran WL. 2007. Skin gammadelta T-cell functions in homeostasis and wound healing. *Immunological reviews* 215:114-22.

Khanna S, Biswas S, Shang Y, Collard E, Azad A, Kauh C, Bhasker V, Gordillo GM, Sen CK, Roy S. 2010. Macrophage dysfunction impairs resolution of inflammation in the wounds of diabetic mice. *PloS one* 5:e9539.

Laplante AF, Germain L, Auger FA, Moulin V. 2001. Mechanisms of wound reepithelialization: hints from a tissue-engineered reconstructed skin to long-standing questions. *FASEB journal: official publication of the Federation of American Societies for Experimental Biology* 15:2377-89.

Larjava H, Salo T, Haapasalmi K, Kramer RH, Heino J. 1993. Expression of integrins and basement membrane components by wound keratinocytes. *The Journal of clinical investigation* 92:1425-35.

Leung A, Crombleholme TM, Keswani SG. 2012. Fetal wound healing: implications for minimal scar formation. *Current opinion in pediatrics* 24:371-8.

Lucas T, Waisman A, Ranjan R, Roes J, Krieg T, Muller W, Roers A, Eming SA. 2010. Differential roles of macrophages in diverse phases of skin repair. *Journal of immunology* 184:3964-77.

Macleod AS, Havran WL. 2011. Functions of skin-resident gammadelta T cells. *Cellular and molecular life sciences : CMLS* 68:2399-408.

Martin P, D'Souza D, Martin J, Grose R, Cooper L, Maki R, McKercher SR. 2003. Wound healing in the PU.1 null mouse--tissue repair is not dependent on inflammatory cells. *Current biology : CB* 13:1122-8.

Martinotti, S, Burlando, B.; Ranzato, E. 2012. Role of extracellular matrix in wound repair process. In Maria Eduarda Henriques and Marcio Pinto editors. Type I Collagen: *Biological Functions, Synthesis and Medicinal Applications Process. Hauppauge*, New York: Nova Publishers Inc.

Meszaros AJ, Reichner JS, Albina JE. 2000. Macrophage-induced neutrophil apoptosis. *Journal of immunology* 165:435-41.

Mills RE, Taylor KR, Podshivalova K, McKay DB, Jameson JM. 2008. Defects in skin gamma delta T cell function contribute to delayed wound repair in rapamycin-treated mice. *Journal of immunology* 181:3974-83.

Monaco JL, Lawrence WT. 2003. Acute wound healing an overview. *Clinics in plastic surgery* 30:1-12.

Montesano R, Orci L. 1988. Transforming growth factor beta stimulates collagen-matrix contraction by fibroblasts: implications for wound healing. *Proceedings of the National Academy of Sciences of the United States of America* 85:4894-7.

Nirodi CS, Devalaraja R, Nanney LB, Arrindell S, Russell S, Trupin J, Richmond A. 2000. Chemokine and chemokine receptor expression in keloid and normal fibroblasts. Wound repair and regeneration: *official publication of the Wound Healing Society [and] the European Tissue Repair Society* 8:371-82.

Nissen NN, Polverini PJ, Koch AE, Volin MV, Gamelli RL, DiPietro LA. 1998. Vascular endothelial growth factor mediates angiogenic activity during the proliferative phase of wound healing. *The American journal of pathology* 152:1445-52.

Parks WC. 1999. Matrix metalloproteinases in repair. Wound repair and regeneration : *official publication of the Wound Healing Society [and] the European Tissue Repair Society* 7:423-32.

Pilcher BK, Wang M, Qin XJ, Parks WC, Senior RM, Welgus HG. 1999. Role of matrix metalloproteinases and their inhibition in cutaneous wound healing and allergic contact hypersensitivity. *Annals of the New York Academy of Sciences* 878:12-24.

Raja, Sivamani K, Garcia MS, Isseroff RR. 2007. Wound re-epithelialization: modulating keratinocyte migration in wound healing. *Frontiers in bioscience : a journal and virtual library* 12:2849-68.

Ranzato E, Burlando B. 2011. Signaling pathways in wound repair. In Middleton J, editor. *Wound healing: process, phases and promoting.* Hauppauge, New York: Nova Publishers Inc, p 123-135.

Ranzato E, Martinotti S, Volante A, Mazzucco L, Burlando B. 2011. Platelet lysate modulates MMP-2 and MMP-9 expression, matrix deposition and cell-to-matrix adhesion in keratinocytes and fibroblasts. *Experimental dermatology* 20:308-13.

Ranzato E, Mazzucco L, Burlando B. 2010. Platelet Derivatives: A New Horizon in Regenerative Medicine. In Berhardt LV, editor. Advances in *Medicine and Biology.* Hauppauge, New York: Nova Publishers Inc, p 245-254.

Remensnyder JP, Majno G. 1968. Oxygen gradients in healing wounds. *The American journal of pathology* 52:301-23.

Rodero MP, Khosrotehrani K. 2010. Skin wound healing modulation by macrophages. *International journal of clinical and experimental pathology* 3:643-53.

Rosen A, Casciola-Rosen L. 1999. Autoantigens as substrates for apoptotic proteases: implications for the pathogenesis of systemic autoimmune disease. *Cell death and differentiation* 6:6-12.

Serini G, Valdembri D, Bussolino F. 2006. Integrins and angiogenesis: a sticky business. *Experimental cell research* 312:651-8.

Singer AJ, Clark RA. 1999. Cutaneous wound healing. *The New England journal of medicine* 341:738-46.

Stoler A, Duvic M, Fuchs E. 1989. Unusual patterns of keratin expression in the overlying epidermis of patients with dermatofibromas: biochemical alterations in the epidermis as a consequence of dermal tumors. *The Journal of investigative dermatology* 93:728-38.

Tonnesen MG, Feng X, Clark RA. 2000. Angiogenesis in wound healing. The journal of investigative dermatology. *Symposium proceedings / the Society for Investigative Dermatology, Inc. [and] European Society for Dermatological Research* 5:40-6.

Werner S, Grose R. 2003. Regulation of wound healing by growth factors and cytokines. *Physiological reviews* 83:835-70.

Xue M, Le NT, Jackson CJ. 2006. Targeting matrix metalloproteases to improve cutaneous wound healing. *Expert opinion on therapeutic targets* 10:143-55.

In: Keratinocytes
Editor: Elia Ranzato

ISBN: 978-1-62618-798-6
© 2013 Nova Science Publishers, Inc.

Chapter 2

Regulation of the Proliferation and Differentiation of Keratinocytes and Cellular Interaction between Keratinocytes and the Tissue Environment

Tomohisa Hirobe[*]
Fukushima Restoration Support Headquarters,
National Institute of Radiological Sciences,
Anagawa, Inage-ku, Chiba, Japan

Abstract

Keratinocytes comprise the bulk of the epithelium, undergo keratinization and form dead superficial layers of the skin. These superficial keratinized cells are continuously replaced by cells derived from mitotic cells in the lowest layer of the epidermis (basal layer). The proliferation and differentiation of keratinocytes is regulated by many factors including hormones, growth factors and cytokines. Melanocytes are cells in the basal layers of the epidermis that do not keratinize but can

[*] Phone: +81-43-206-3253, Fax: +81-43-206-4638, E-mail: thirobe@nirs.go.jp.

produce melanin pigments. Keratinocytes and melanocytes migrate from the epidermis to the dermis to form pigmented hairs. Mammalian hair is formed by three kinds of cell, namely keratinocytes and melanocytes derived from the epidermis and fibroblasts consisting of the dermal papilla. Cellular interactions between keratinocytes and melanocytes/fibroblasts are important for the development and maintenance of hair. Keratinocyte stem cells and melanocyte stem cells locate in the bulge of hair follicles and produce new proliferating and differentiating keratinocytes and melanocytes at the new hair growth cycle. The proliferation and differentiation of keratinocytes appear to be regulated by their own autocrine factors as well as by paracrine factors derived from melanocytes and fibroblasts. Keratinocytes also produce and release proliferation- and differentiation-stimulating factors towards melanocytes and fibroblasts. Thus, in addition to the autocrine factors, the paracrine factors derived from the tissue environment such as melanocytes and fibroblasts are thought to play an important role in the regulation of keratinocyte function as well as maintenance of skin homeostasis.

Introduction

The skin is the largest organ in the human body. The skin covers the surface of the body and consists of three main layers; the surface epidermis, subjacent dermis and the lowest layer, subcutaneous layer. The skin possesses many functions including protection of the human body from the invasion of chemical substances, bacteria and viruses by keratinization of the epidermis (barrier); control of body temperature by expansion and contraction of blood vessels or capillaries; detection of external stimuli and its signal transmission to the brain (sensory organ); absorption of oily or other substances from the epidermis; secretion of sweat or sebum from the sweat gland or sebaceous gland; protection of the skin from invasion of alien substances using cellular immunity; oxygen is incorporated into capillaries abundantly distributed in the skin (respiration) and protection of the internal organs from physical stimuli (the integrity of the skin is supported by collagen and elastin fibers in the dermis). These skin functions are supported by various cells including epithelial tissue cells such as keratinocytes and sebaceous gland cells, connective tissue cells such as fibroblasts and endothelial cells, immune system cells such as T cells, leucocytes and Langerhans cells and ectoderm-derived cells such as melanocytes and Merkel cells (Hirobe, 2005; Tobin, 2006).

Keratinocytes comprise the bulk of the epithelium, undergo keratinization and form the dead superficial layers of the skin. These superficial keratinized cells continuously desquamate from the surface and are replaced by cells derived from mitotic cells in the lowest layer of the epidermis (basal layer). The cells are displaced to successively higher levels by the population of new cells below them. As they move upwards, they elaborate keratin and accumulate in the cytoplasm, and finally cells are mostly occupied by keratin. In addition to keratinocytes, melanocytes are important cells constituting the epidermis. Melanocytes are cells in the basal layers that do not keratinize but can produce melanin pigments (Figure 1A; Quevedo and Smith 1963; Szabo, 1967; Takeuchi, 1985; Hirobe, 1995). Keratinocytes and melanocytes migrate into the dermis to form pigmented hair (Figure 1B; Hirobe, 1984, 1992a).

Mammalian hair consists of keratinocytes and melanocytes derived from the epidermis in addition to fibroblasts (constitute the dermal papilla) derived from the dermis (Chase, 1954; Hirobe, 1995; Peters, 2002). Keratinocyte stem cells (Kobayashi et al., 1993; Cotsarelis, 2006a, b) and melanocyte stem cells (Nishimura et al., 2002, 2005, 2010; Inomata et al., 2009) locate in the bulge of hair follicles, and they produce new proliferating and differentiating keratinocytes and melanocytes at the next hair growth cycle (Osawa et al., 2005; Nishikawa-Torikai et al., 2011). The proliferation and differentiation of keratinocytes are regulated by many growth factors and cytokines. Keratinocytes are also known to regulate the proliferation and differentiation of mammalian epidermal melanocytes by keratinocyte-derived mitogens and melanogens. Moreover, keratinocytes produce and release mitogenic factors towards fibroblasts.

In contrast, fibroblasts produce and release growth factors and cytokines that regulate the proliferation and differentiation of keratinocytes. The three types of skin cell are mutually interacted for their proliferation and differentiation. In this chapter, in addition to the autocrine regulation of keratinocytes, paracrine regulation by melanocytes and fibroblasts is reviewed and the regulation of skin homeostasis by the cellular interaction is also discussed.

Structure and Function of the Epidermis

Important role of the epidermis is to protect the skin from many kinds of environmental stress such as bacteria, viruses, chemicals, ultraviolet radiations (UV), ionizing radiations, several kinds of electromagnetic waves as well as physical, thermal and mechanical injuries (barrier function). To support this role, the epidermis forms many kinds of appendages such as hair follicles, sebaceous glands and sweat glands.

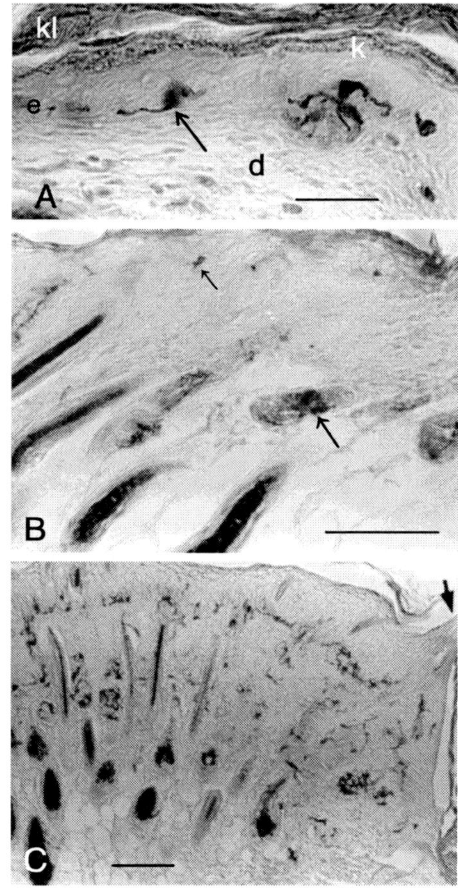

Figure 1. Vertical sections of the dorsal skin of 3.5 (A), 4.5 (B) and 5.5 (C)-day-old C57BL/10J mice. (A) Dopa-positive melanocytes (arrow) are seen in the basal layer of the epidermis (e). (B) Many hair follicles are developed. In addition to epidermal melanocytes (small arrow), melanocytes (arrow) are observed in the hair bulbs. (C) A full-thickness cut 7 mm long was made on the mid-dorsal skin of 1.5-day-old C57BL/10J mice and the skin was fixed on the fourth day after wounding (5.5-day-old mice). Specimens were treated with dopa reagent. Epidermal keratinocytes and melanocytes are proliferated in the vicinity of the cut edge (large arrow). Numerous dopa-positive melanocytes are observed in the epidermis, dermis and hair follicles near the cut edge as well as behind the advancing epidermal sheets. kl, keratinized layer, k, keratohyalin granules; d, dermis. Scale bar: A, 20 μm; B and C, 100 μm.

The epidermis is required to renew constantly from birth to death of the individual to support the maintenance of skin homeostasis and regeneration after damage by many kinds of environmental stress mentioned above. The renewal of the epidermis, normally requires approximately 28 days, and is supported by the proliferation and differentiation of keratinocytes, main cell component of the epidermis. Thus, the epidermal homeostasis mainly depends on a balance between the proliferation and differentiation of keratinocytes.

The epidermis is a morphologically stratified squamous epithelium consisting of mainly of cells with two different origins: ectoderm-derived keratinocytes (Figure 2A) and neural crest-derived melanoytes (Figure 2B).

Keratinocytes comprise the bulk of the epidermis, undergo differentiation (keratinization) and form the dead superficial layers (keratinized layer or cornified layer; Figure 1A) of the skin. These superficial keratinized cells continuously desquamate from the surface and are replaced by cells in the lowest layer (basal layer).

The cells are displaced to successively higher levels by the population of new cells below them. As they move upward, they elaborate keratin and accumulate in the cytoplasm (tonofilament in the basal layer and spinous layer; Figure 2A and keratohyalin granules in the granular layer; Figure 1A), and finally almost all cells are occupied by keratin (keratinized layer or cornified layer; Figure 1A). Although the morphological changes associated with the epidermal differentiation are well studied, the molecular mechanisms underlining this process are still not fully understood.

Mammalian melanocytes (Figure 1A) are cells in the basal layers that do not keratinize but produce melanin pigments (Ito, 2003). These melanin pigments are produced from L-tyrosine (L-Tyr) with the aid of enzymatic reactions performed by tyrosinase, tyrosinase-related protein (TRP)-1 and TRP-2 (Hearing, 2000). Most of these melanocytes migrate from the epidermis to the hair follicles (Figure 1B), and colonize hair bulb melanocytes in hairy general body (trunk) skin. In the glabrous skin of ear, nose, foot and tail of animals as well as in the human skin except scalp, underarm and pubic skin, numerous differentiated melanocytes are present in the epidermis even in the adults (Szabo, 1967; Quevedo and Fleischmann, 1980, Hirobe, 1992a). However, in the epidermis of mouse hairy skin, epidermal melanocytes are found only during the early weeks after birth (Hirobe, 1984). The process of melanin synthesis in mice is regulated by many coat-color genes (Hirobe and Abe, 1999; Hirobe, 2011a). In human skin, the epidermal melanin unit that is composed with keratinocytes and melanocytes plays an important role in the regulation of epidermal pigmentation and in the maintenance of epidermal homeostasis (Szabo, 1967).

Epidermal structure can be regenerated after skin wounding. The epidermal keratinocytes and melanocytes/melanoblasts in neonatal C57BL/10J mouse skin are stimulated to undergo mitosis immediately adjacent to a skin wound, and thereafter to migrate in the regenerating wound epidermis (Figure 1C; Hirobe, 1983, 1988a, b). The density of the melanocyte population in the regenerating epidermis does not differ from that initially (Hirobe, 1983). Epidermal keratinocytes and melanocytes/melanoblasts seem to be able to regenerate to recover normal epidermal structure.

Structure and Function of the Hair

The hair is formed in hair follicles derived from hair germs that begin as an epidermal invagination into the dermis. The dermis forms a thickening beneath the epidermis and the end of the invagination comes to surround it. The dermal thickening develops into the dermal papilla consisting of fibroblasts (Figure 2C), and the surrounding part of the invagination forms the hair bulb (Hirobe, 1992a). The hair follicular melanocytes derived from epidermal melanocytes are highly dendritic cells and colonize the hair matrix, lower half of the hair bulb (Figure 1B, C). Hair bulb melanocytes secrete mature stage IV melanosomes (melanin-containing organelles, Seiji et al., 1963) into keratinocytes consisting of the hair bulb. Keratinocytes develop hair cortex and medulla where melanosomes are incorporated. In mice, the process of morphogenesis of hair follicle is cyclic (Dry, 1926) and called hair growth cycle (Chase, 1954). The hair growth cycle consists of three stages: resting (telogen); growth (anagen); and regression (catagen). Anagen hair follicles produce hair shafts that are formed by fully keratinized cells and pigmented melanocytes (Dry, 1926; Chase, 1954). This deposition of stage IV melanosomes (Fitzpatrick et al., 1969) continues during the entire growth phase of the hair (ca. 17 days). This phase is divided into six substages (anagen I–VI: Chase, 1954). When cell proliferation of hair bulb keratinocytes ceases (catagen), the melanocytes also cease to produce melanosomes and no further cells enter the hair shafts followed by the resting stage (telogen). In mice, hair growth is synchronized and proceeds in waves all over the body (Dry, 1926).

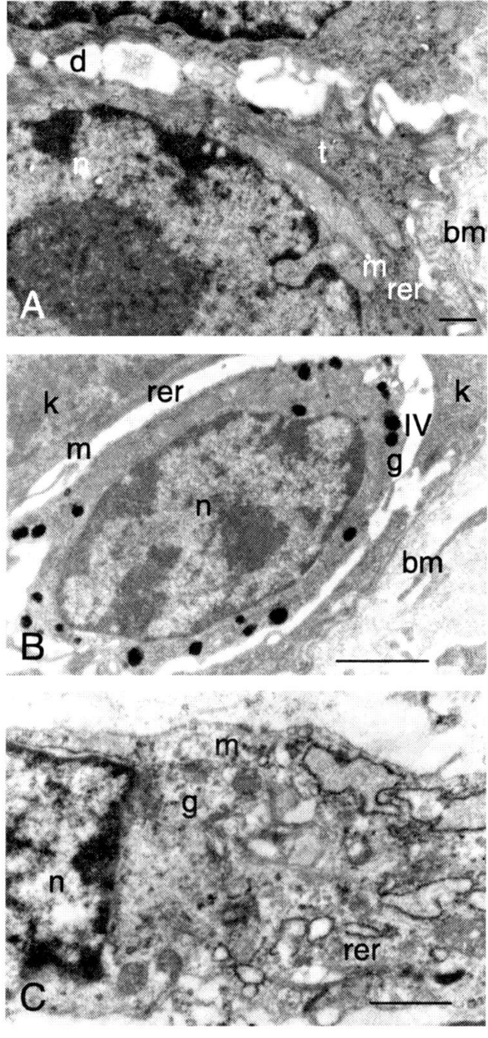

Figure 2. Electron micrographs of an epidermal keratinocytes (A), an epidermal melanocyte (B) and a dermal fibroblast (C) of 3.5-day-old C57BL/10J mice. (A) Many mitochondria (m), rough endoplasmic reticulum (rer) and tonofilament (t) are observed. Basal keratinocytes sit on the basement membrane (bm) and adhere each other by cell junction, desmosome (d). (B) An epidermal melanocyte resides in the basal layer of the epidermis and contacts with basal keratinocytes (k) through small processes. In addition to many mitochondria (m), the Golgi apparatus (g) and rough endoplasmic reticulum (rer) and many mature stage IV (IV) melanosomes are seen. (C) A dermal fibroblast is seen under the epidermis. Many mitochondria (m), the Golgi apparatus (g) and rough endoplasmic reticulum (rer) are seen in the cytoplasm. n: nucleus, Scale bar: 1 μm.

Hair bulb melanocytes differ from epidermal melanocytes in that hair bulb melanocytes are larger and possess larger dendrites and interact with fewer keratinocytes than epidermal melanocytes (Hirobe, 1995). In mature hair follicles, a functional melanin unit between melanocytes and keratinocytes in the hair bulb is formed. Hair bulb melanocytes locate in the basal layers of the hair matrix in close proximity to the dermal papilla consisting of numerous fibroblasts and rest on the basement membrane. Melanogenesis is strictly coupled to the growth phase of the hair growth cycle (anagen III-VI, Chase, 1954). Melanogenesis ceases early in catagen and is completely absent in telogen (Chase, 1954). Towards the end of anagen the number of identifiable melanocytes decreases and melanocytes lose their dendrites, shrink and become less pigmented, then they disappear in catagen (Chase, 1954). Keratinocytes and melanocytes die and finally hair follicles move upwards to form rudimentary hair germ near the sebaceous gland. Keratinocyte stem cells and melanocyte stem cells are present in the bulge of hair follicles in telogen stage (Nishimura et al., 2002).

It has been reported that telogen hair follicles can be induced to enter new anagen by plucking hair shafts. The method induces the highly synchronized development of homogeneous anagen hair follicle populations over the entire depilated skin area. Silver et al. (1969) observed the depilation-induced anagen on epon-embedded, semi-thin sections of mouse skin of C57BL/Ch, BUB/Wi, C3HP/Wi, DBA/1Ch and DS/Ch strains. A population of non-dividing (clear dendritic cells) of the hair germ immediately adjacent to the dermal papilla represented the melanoblasts (or melanocyte stem cells) of telogen hair follicles, which went on to produce melanin in anagen. These clear dendritic cells underwent mitosis before they resumed melanogenesis during anagen III-IV, and changed back into dormancy during telogen. Molecular characteristics of melanogenesis during the depilation-induced hair growth cycle in mice have been studied. The levels of tyrosinase mRNA and protein as well as the melanosomal protein gp75 were not detectable in telogen of C57BL/6 mice (Slominski et al., 1991). During the 1st and 2nd days of anagen melanin was absent, and gp75 and tyrosinase activity were also undetectable (Slominski et al., 1991). However, within 1 day of anagen induction, tyrosinase mRNA and protein were around the detectable level, and were clearly identifiable by the 2nd day. Sugiyama (1979) also demonstrated by electron microscopic observations the presence of poorly differentiated melanocytes containing unmelanized, but dopa-positive melanosomes in early anagen. The abundance of tyrosinase mRNA, protein and activity as well as gp75 increased rapidly by the 5th day, reaching a plateau by the 8th-12th days (Slominski et al., 1991).

Melanin was first detected by the 4th-5th days of anagen and become abundant by the 8th-12th days (Slominski et al., 1991; Slominski and Paus, 1993). Burchill et al. (1989) also demonstrated that in C3H/He-A^{vy} (viable yellow) mouse skin, tyrosinase activity was dependent both on enzyme synthesis and post-translational modification during depilation-induced anagen. The activity of TRP-2 similarly increased at the early and middle anagen, and decreased at the late anagen of depilation-induced hair growth cycle of C57BL/6 mice (Slominski et al., 1994). However, the increase in TRP-2 activity was much smaller than that of tyrosinase activity, suggesting that tyrosinase is the major regulator of melanogenesis in the hair growth cycle, whereas TRP-2 plays a modulatory role.

The mechanisms of the initiation of hair growth cycle are still unknown. Paus et al. (1994) reported that anagen-associated decline in the number of mast cells of C57BL/6 mice was correlated with the occurrence of substantial mast cell degranulation and the secretory products from the mast cell granules induced the development of new anagen follicles. These studies suggest that mast cells play an important role in the initiation of mouse hair growth cycle. Recently, β-catenin is reported to regulate the differentiated function of hair bulb melanocytes (Enshell-Seijffers, 2010).

Proliferation and Differentiation of Keratinocytes

The proliferation and differentiation of keratinocytes is involved in regulating the function of keratinocytes throughout skin development. The proliferation and differentiation of keratinocytes are regulated by hormones and autocrine and paracrine factors such as epidermal growth factor (EGF; Rheinwald and Green, 1977), keratinocyte growth factor (KGF; Marchese et al., 1989), hydrocortisone (Hirobe, 1994) and dexamethasone (Hirobe, 1994). Proliferating and differentiating keratinocytes are derived from keratinocyte stem cells in both epidermis and hair follicles. Keratinocyte stem cells are responsible for maintaining the epidermal homeostasis, hair growth cycle and regeneration after skin wounding. Keratinocyte stem cells perform self-renewal and produce differentiated lineages that form completed tissues, namely epidermis and hair follicles. It is known that three populations of keratinocyte stem cells exist in different skin areas, namely interfollicular stem cells (epidermal stem cells) in the basal layer, the bulge of hair follicles (hair follicle stem cells) and sebaceous glands (sebaceous gland stem cells). During the renewal and regeneration after skin wounding, the three kinds of keratinocyte stem cell seem to be maintained by their own stem cells. However, at the time of disruption of skin homeostasis, all the populations of the three keratinocyte stem cells seem to be able to reproduce all the three structures (Levy et al., 2007).

In the interfollicular epidermis, the regeneration is performed by the proliferation of epidermal stem cells, subpopulation of basal keratinocytes. The epidermal stem cells can divide infrequently to produce daughter stem cells and their progenitor cells that are named transit amplifying cells. Keratinocyte stem cells locate in a special microenvinronment named niche that allows them to maintain their characteristics and stemness. Epidermal stem cells are known to locate in the basal layer and rest on the basement membrane that is rich in intracellular matrix proteins and growth factors. According to the exit from the niche, basal keratinocytes move to suprabasal layers, and different microenvironment can stimulate their destiny.

In hair follicles, the new hair is regenerated from hair follicle-specific stem cells reside in the hair bulge (Kobayashi et al., 1993). The hair bulge provides a unique differentiation-restricted environment for different types of adult stem cells. In fact, the hair bulge contains melanocyte stem cells (Nishimura et al, 2002) with the characteristics of neural crest cells (Yu et al., 2010). Keratinocyte stem cells in the hair follicles of mouse dorsal skin and vibrissa are identified as quiescent cells in vivo by label-retaining studies (Cotsarelis et al., 1990). Hair bulge cells in adult mice were then shown through genetic labeling studies to generate all the cell lineages within hair follicles and to function as stem cells after isolation (Morris, 2000).

By transplantation of human scalp tissue to immunodeficient nude mice, label-retaining cells in human hair follicles were shown to localize in the basal layer of the outer root sheath layer of the hair bulge (Lyle et al., 1998). In contrast to murine hair bulge, human hair bulge is not morphologically apparent. Therefore, markers for keratinocyte stem cells have been searched for many years. Human hair bulge express CD200, K15 and K19, while murine hair bulge cells express CD34 and K15 (Pincelli and Marconi, 2010).

The third keratinocyte stem cells are identified in the sebaceous gland that arises from the hair follicles. Biomarkers for the sebaceous gland stem cells are Blimp-1, and conditional ablation of Blimp-1 induces sebaceous gland hyperplasia. Lineage tracing experiments confirm that the sebaceous gland stem cells can give rise to the entire gland, thus ensuring sebaceous gland homeostasis (Horsley et al., 2006).

Interaction with Melanocytes

Many of important findings concerning the role of keratinocytes in the regulation of the proliferation and differentiation of mammalian melanocytes are derived from in vitro study. Serum-free primary culture system of melanoblasts/melanocytes and keratinocytes is preferable because unidentified factors present in the serum can be eliminated.

Pure and enriched cultures of primary mouse keratinocytes were obtained by culturing epidermal cell suspensions of C57BL/10 J mice with keratinocyte-defined medium (KDM) consisting of Ca^{2+}-free minimum essential medium (MEM) supplemented with MEM-nonessential amino acid solution, insulin (Ins), bovine serum albumin (BSA), ethanolamine (EA), phosphoethanolamine (PEA), sodium selenite (SE), epidermal growth factor (EGF), hydrocortisone (HC), dexamethasone (Dex) and 0.03 mM $CaCl_2$ (Hirobe, 1992b). The primary keratinocytes can be trypsinized and seeded into the pure melanoblasts obtained in melanoblast-defined medium (MDM), consisting of Ham's F-10 medium supplemented with Ins, BSA, EA, PEA and SE at 14 days (Hirobe, 1992c). Alpha-melanocyte-stimulating hormone (MSH), adrenocorticotrophic hormonre (ACTH) or 3-isobutyl-1-methylxanthine (IBMX) induced the differentiation of melanoblasts into melanocytes in the presence of keratinocytes but not in the absence of keratinocytes (Hirobe, 1992c; Hirobe and Abe, 2000). Keratinocyte-derived melanogens have also been revealed to be endothelin (ET)-1 (Hirobe, 2001), ET-2 (Hirobe, 2001), ET-3 (Hirobe, 2001), leukemia inhibitory factor (LIF; Hirobe, 2002), stem cell factor (SCF; Hirobe et al., 2003), hepatocyte growth factor (HGF; Hirobe et al., 2004a) and granulocyte-macrophage colony-stimulating factor (GMCSF; Hirobe et al., 2004b, c).

In animals, α-MSH is also produced in the skin of rat (Thody et al., 1983) and gerbil (Thody et al., 1983). In guinea pig, SCF stimulates the proliferation and melanogenesis of melanocytes in UVB-irradiated skin (Hachiya et al., 2001). In mice, α-MSH is produced in the skin of adult hairless mouse (Thody et al., 1983). Alpha-MSH is also produced and released from SP-1 keratinocyte cell line and stimulates the melanogenesis of melan-a cells (Virador et al., 2001).

However, the gene for proopiomelanocortin (POMC) was not expressed in cultured keratinocytes of newborn mice as well as in the epidermis and dermis separated by trypsin from prenatal and postnatal mice (Hirobe et al., 2004d), suggesting that no keratinocyte in prenatal and neonatal mouse skin produces and releases α-MSH or ACTH (Hirobe et al., 2004d). In mice, the major source of α-MSH and/or ACTH may be derived from the pituitary through the blood stream, but not from epidermal keratinocytes at least in prenatal and postnatal stages (Hirobe et al., 2004d).

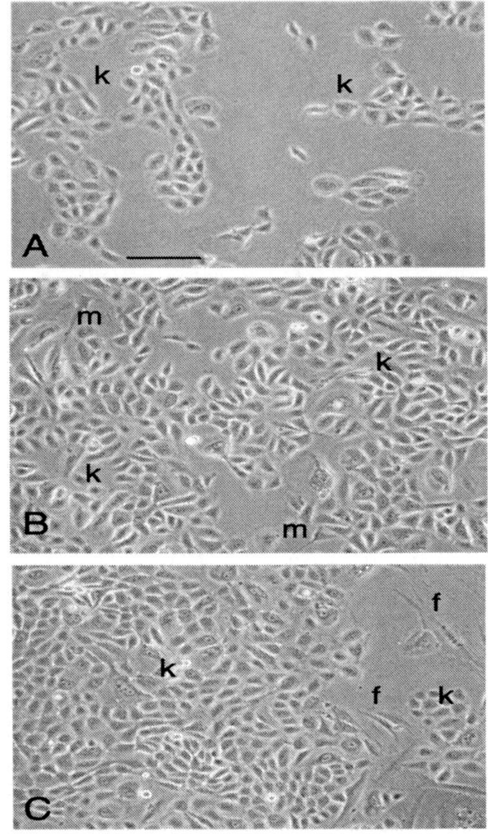

Figure 3. Culture of human epidermal keratinocytes (A), co-culture of keratinocytes and epidermal melanoblasts (B) or dermal fibroblasts (C). Serial cultures of keratinocytes, melanoblasts and fibroblasts derived from newborn foreskins were prepared (Kurabo, Osaka, Japan). Human keratinocytes were cultured with or without melanoblasts or fibroblasts using KG2 medium. After 3 days, keratinocytes (k) proliferated well in the presence of melanoblasts (m; B) or fibroblasts (f; C) compared with keratinocytes alone (A). In addtion to growing keratinocytes, large keratinocytes (differentiating) are also seen in the co-culture of both keratinocytes–melanoblasts and keratinocytes–fibroblasts, suggesting that melanoblasts and fibroblasts produce and release proliferation- and differentiation-stimulating factors towards keratinocytes. Scale bar, 100 μm.

In contrast, DBcAMP supplemented into MDM at high concentrations (MDMD medium) induced the proliferation of melanocytes in the presence of keratinocytes, but not in the absence of keratinocytes (Hirobe, 1992c, 1994), suggesting that keratinocytes produce factors that stimulate the proliferation of mouse epidermal melanocytes in an increased cAMP level. It has been reported that the keratinocyte-derived mitogens for melanocytes are ET-1 (Imokawa et al., 1997; Tada et al., 1998; Hirobe, 2001), ET-2 (Hirobe, 2001), ET-3 (Hirobe, 2001), LIF (Hirobe, 2002), SCF (Kunisada et al., 1998; Hirobe et al., 2003), HGF (Kunisada et al., 2000; Hirobe et al., 2004a) and GMCSF (Imokawa et al., 1996; Hirobe et al., 2004b,c).

Pure and enriched populations of melanoblasts (approximately 90%) and melanocytes (approximately 10%) were obtained by culturing mouse epidermal cell suspensions with MDMDF, consisting of MDMD and basic fibroblast growth factor (bFGF) (Hirobe, 1992b). They were trypsinized and seeded into new dishes.

At 1 day of secondary culture of pure melanoblasts and melanocytes in MDMDF, subconfluent primary keratinocytes in KDM were trypsinized and seeded. The melanoblasts proliferated well in the presence of keratinocytes but not in the absence of keratinocytes (Hirobe, 1992b; Furuya et al., 2002).

These results suggest that keratinocytes produce mitogenic and melanogenic factors towards melanoblasts/melanocytes. These keratinocyte-derived mitogens for melanoblasts have been shown to be ET-1, ET-2, ET-3, LIF, SCF, HGF and GMCSF (Hirobe, 2005).

In the human epidermis, α-MSH (Thody et al., 1983; Schauer et al., 1994; Abdel-Malek et al., 1995; Chakraborty et al., 1996; Wakamatsu et al., 1997; Slominski et al., 2000), ACTH (Chakraborty et al., 1996; Wakamatsu et al., 1997; Slominski et al., 2000)/ACTH fragments (Wakamatsu et al., 1997; Slominski et al., 2000) and nerve growth factor (NGF, Yaar et al., 1991) are produced in and released from keratinocytes and are involved in regulating the melanogenesis and/or dendritogenesis of melanocytes in primary and/or serial culture. ET-1 (Yada et al., 1991; Imokawa et al., 1992; Yohn et al., 1993; Hara et al., 1995) and GMCSF (Imokawa et al., 1996) are the keratinocyte-derived factors that regulate the proliferation and melanogenesis/dendritogenesis of melanocytes in UVB or UVA-irradiated human skins. Prostaglandin (PG) E_2 and PGF_{2a} are produced and released from human keratinocytes by the stimulation of proteinase-activated receptor 2 (PAR-2) and to stimulate the dendritogenesis of human epidermal melanocytes in culture (Scott et al., 2004). By contrast, bFGF (Halaban et al., 1988) is involved in regulating melanocyte proliferation only. SCF is also expressed in cultured keratinocytes

(Hachiya et al., 2001) and regulates the proliferation and differentiation of human melanocytes.

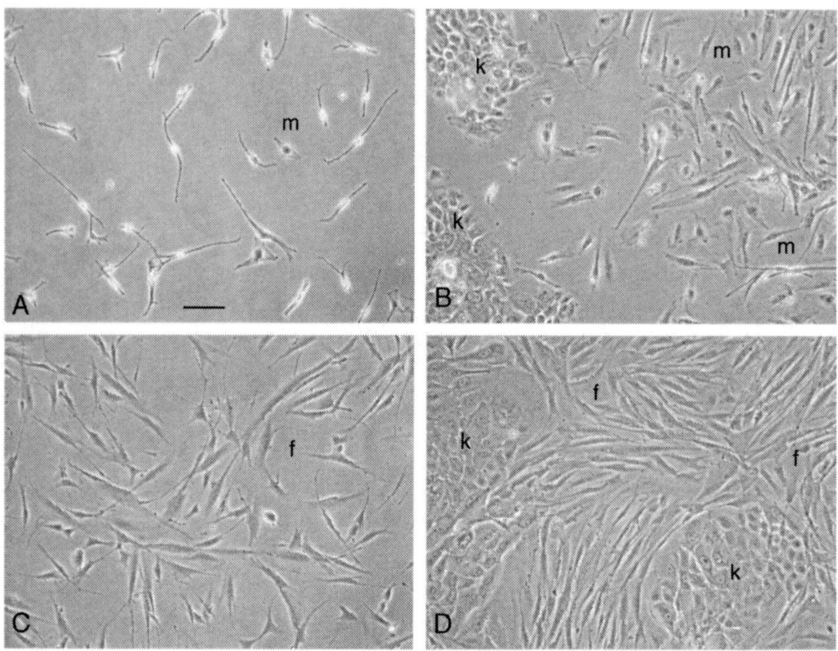

Figure 4. Effects of human epidermal keratinocytes on the proliferation and differentiation of epidermal melanoblasts or dermal fibroblasts. Melanoblasts (m; A) were cultured in MDMD supplemented with Tf, ET-1 and SCF with (B) or without keratinocytes (k). After 3 days in co-culture, melanocytes proliferated well in the presence of keratinocytes (B) compared with melanoblasts/melanocytes alone (A). Numerous differentiating melanocytes with well-developed dendrites and melanin pigments throughout the cytoplasm are seen near keratinocyte (k) colony, suggesting that keratinocytes produce and release proliferation- and differentiation-stimulating factors towards melanoblasts. After 3 days in co-culture with HFDM-1 medium, fibroblasts (f) proliferated well in the presence of keratinocytes (D) compared with fibroblasts alone (C). Many fibroblasts with extended cytoplasm are seen near keratinocyte colonies, suggesting that keratinocytes produce and release proliferation- and differentiation-stimulating factors towards fibroblasts. Scale bar, 100 μm.

These factors are involved in regulating the proliferation and melanogenesis/ dendritogenesis of human epidermal melanocytes in normal skin (Thody et al., 1983; Yaar et al., 1991; Schauer et al., 1994; Wakamatsu et al., 1997) and/or in UVA (Imokawa et al., 1996)/UVB (Halaban et al., 1988;

Yada et al., 1991; Imokawa et al., 1992; Hara et al., 1995; Chakraborty et al., 1996; Hachiya et al., 2001; Imokawa, 2004)-irradiated skin.

Human epidermal melanocytes can also be cultured with MDMD supplemented with transferrin (Tf), ET-1 and SCF, and human epidermal melanoblasts can be cultured with MDMDF supplemented with Tf, ET-1 and SCF (Hirobe, 2009, 2011b, 2012; Hirobe et al., 2010). Human epidermal keratinocytes co-cultured with human epidermal melanocytes (Figure 3B) proliferated better than keratinocytes alone in KG2 medium (Kurabo, Osaka, Japan; Figure 3A), suggesting that human epidermal melanocytes produce and release proliferation-stimulating factors towards keratinocytes. Moreover, many keratinocytes (Figure 3B) co-cultured with melanocytes differentiated (possessed enlarged cytoplasm) better than keratinocytes alone (Figure 3A), suggesting that human epidermal melanocytes produce and release differentiation-stimulating factors towards keratinocytes. In contrast, human epidermal melanoblasts (Figure 4B) co-cultured in MDMD supplemented with Tf, ET-1 and SCF (Hirobe, 2009, 2011b, 2012; Hirobe et al., 2010) with human epidermal keratinocytes proliferated and differentiated much better than melanoblasts alone (Figure 4A). Numerous dendritic and pigmented melanocytes were observed near keratinocyte colonies (Figure 4B). These results suggest that human epidermal keratinocytes produce and release melanocyte mitogens and melanogens. Taken together, human and animal keratinocytes produce and release melanocyte mitogens and melanogens in a similar fashion.

Interaction with Fibroblasts

Epithelial-mesenchymal interactions play an important role in the regulation of homeostasis and repair of several tissues in mammals. In the skin, the interaction of epidermal keratinocytes and dermal fibroblasts is important for the normal development and regeneration after wounding. Epidermal keratinocytes are separated from dermal mesenchymal cells by the basement membrane, which is composed of complex extracellular matrix such as type I and IV collagens, rendering direct cell-cell contact-mediated mechanisms are less probable in situ. Thus, the interaction between keratinocytes and fibroblasts is mainly performed by soluble factors displaying paracrine activity and by cell-matrix interactions. Using co-cultures of normal human epidermal keratinocytes and dermal fibroblasts, Maas-Szabowski et al. (1999) found that keratinocytes co-cultured with fibroblasts stimulated the production and release of interleukin (IL)-1α and IL-1β, and IL-1α and IL-1β induced the production and release of KGF in fibroblasts. KGF secreted from fibroblasts stimulated the proliferation of keratinocytes in culture. Thus, the regulation of the proliferation of keratinocytes may be performed via a paracrine mechanism (Maas-Szabowski et al., 1999). Moreover, keratinocytes produce and release mitogenic factors towards fibroblasts such as bFGF (Tamm et al., 1991), acidic FGF (aFGF, Tamm et al., 1991), transforming growth factor-β_1 (TGF-β_1, Maas-Szabowski et al., 1999), KGF (Maas-Szabowski et al., 1999) and platelet-derived growth factor (PDGF, Li et al., 2004). In contrast, fibroblasts produce and release growth factors and cytokines that regulate the proliferation and differentiation of keratinocytes, such as bFGF (Tamm et al., 1991) and PDGF (Li et al., 2004) in addition to KGF.

Human epidermal keratinocytes co-cultured with human dermal fibroblasts (Figure 3C) using KG2 medium proliferated better than keratinocytes alone (Figure 3A), suggesting that human dermal fibroblasts produce and release proliferation-stimulating factors towards keratinocytes. Moreover, many keratinocytes co-cultured with fibroblasts differentiated (Figure 3C, possessed enlarged cytoplasm) better than keratinocytes alone (Figure 3A), suggesting that human dermal fibroblasts produce and release differentiation-stimulating factors towards keratinocytes. In contrast, human dermal fibroblasts co-cultured with human epidermal keatinocytes (Figure 4D) using HFDM-1 medium (IFP, Higashine, Japan) proliferated and differentiated much better than fibroblasts alone (Figure 4C). Numerous fibroblasts with enlarged cytoplasm were observed near keratinocyte colonies (Figure 4C). These results suggest that human epidermal keratinocytes produce and release proliferation- and differentiation-stimulating factors towards fibroblasts and, in addition, human dermal fibroblasts produce and release proliferation- and differentiation-stimulating factors towards keratinocytes.

In addition to the regulation by the paracrine soluble factors, cell-matrix interaction is important for the interaction of keratinocytes and fibroblasts. Since mammalian epidermis and dermis are connected with basement membrane, the function of basement membrane is very important for the interaction between epidermis and dermis. Heparan sulfate (HS) chains of perlecan at the basement membrane act as a reservoir of heparin-binding growth factors such as FGFs (Patel et al., 2007), HGF (Deakin et al., 2009), GMCSF (Sebollela et al., 2005), neuregulin (Pankonin et al., 2005) and vascular endothelial growth factor-A (VEGF-A) (Ashikari-Hada et al., 2005), and serve to prevent uncontrolled diffusion of these factors from the epidermis to the dermis and vice versa (Whitelock et al., 2005). HGF (Mildner et al., 2007; Kovacs et al., 2010), KGF (Maas-Szabowski et al., 1999; Cardinali et al., 2008; Kovacs et al., 2010) and SLF (Kovacs et al., 2010) are produced by dermal fibroblasts and act on epidermal keratinocytes and/or melanocytes. In contrast, VEGF-A is synthesized in the epidermis and promotes vascular endothelial cell growth in the dermis (Detmar et al., 1995). Heparanase is activated in the UVB-irradiated skin, and this leads to loss of heparan sulfate at the basemenet membrane, resulting in uncontrolled diffusion of heparan sulfate-binding cytokines through the basement membrane. This loss of heparan sulfate at the basement membrane is also involved in the pigmentation process in UVB-exposed skin (Iriyama et al., 2011).

Conclusion

Keratinocytes comprise the bulk of the epithelium, undergo keratinization and form the dead superficial layers of the skin. Melanocytes are cells in the basal layers of the epidermis that do not keratinize but can produce melanin pigments. Melanocytes and keratinocytes comprise the epidermal melanin unit in the epidermis, and, in addition, keratinocytes and melanocytes migrate into the dermis to form pigmented hair. Mammalian hair consists of fibroblasts present in the dermal papilla in addition to keratinocytes and melanocytes derived from the epidermis. Keratinocyte stem cells and melanocyte stem cells locate in the bulge of hair follicles, and they produce new proliferating and differentiating keratinocytes and melanocytes at the new hair growth cycle. The proliferation and differentiation of keratinocytes are regulated by their own autocrine factors as well as paracrine factors derived from melanocytes and fibroblasts. Keratinocytes also produce and release proliferation- and differentiation-stimulating factors towards melanocytes and fibroblasts. Thus, the double paracrine regulation of the proliferation and differentiation between keratinocytes and melanocytes /fibroblasts seems to play an important role in the maintenance of skin homeostasis in mammals.

References

Abdel-Malek, Z., Swope, V. B., Suzuki, I., Akcali, C., Harriger, M. D., Boyce, S.T., Urabe, K., and Hearing, V.J. (1995). Mitogenic and melanogenic stimulation of normal human melanocytes by melanotropic peptides. *Proc. Natl. Acad. Sci. USA* 92, 1789–1793.

Ashikari-Hada, S., Habuchi, H., Kariya, Y., Itoh, N., Reddi, A. H., and Kimata, K. (2004). Characterization of growth factor-binding structures in heparin/heparan sulfate using an octasaccharide library. *J. Biol. Chem.* 279, 12346–12354.

Burchill, S. A., Ito, S., and Thody, A. J. (1989). Regulation of tyrosinase synthesis and its processing in the hair follicular melanocytes of the mouse during eumelanogenesis and pheomelanogenesis. *J. Invest. Dermatol.* 93, 236–240.

Cardinali, G., Ceccarelli, S., Kovacs, D., Aspite, N., Lotti, L. V., Torrisi, M. R., and Picardo, M (2005). Keratinocyte growth factor promotes melanosome transfer to keratinocytes. *J. Invest. Dermatol.* 125, 1190–1199.

Chakraborty, A. K., Funasaka, Y., Slominski, A., Ermak, G., Hwang, J., Pawelek, J. M., and Ichihashi, M. (1996). Production and release of proopiomelanocortin (POMC) derived peptides by human melanocytes and keratinocytes in culture:regulation by ultraviolet *B. Biochim. Biophys. Acta* 1313, 130–138.

Chase, H. B. (1954). Growth of the hair. *Physiol. Rev.* 34, 113–126.

Cotsarelis, G. (2006a). Epithelial stem cells: a folliculocentric view. *J. Invest. Dermatol.* 126, 1459–1468.

Cotsarelis G. (2006b). Gene expression profiling gets to the root of human hair follicle stem cells. *J. Clin. Invest.* 116, 19–22.

Cotsarelis, G., Sun, T. T., and Lavker, R. M. (1990). Label-retaining cells reside in the bulge area of pilosebaceous unit: implications for follicular stem cells, hair cycle, and skin carcinogenesis. *Cell* 61, 1329–1337.

Deakin, J. A., Blaum, B. S., Gallagher, J. T., Uhrin, D., and Lyon, M. (2009). The binding properties of minimal oligosaccharides reveal a common heparan sulfate/dermata sulfate-binding site in hepatocyte growth factor/scatter factor that can accommodate a wide variety of sulfation patterns. *J. Biol. Chem.* 284, 6311–6321.

Detmar, M., Yeo, K.-T., Nagy, J. A., van de Water, L., Brown, L. F., Berse, B., Elicker, B. M., Ledbetter, S., and Dvorak, H. F. (1995). Keratinocyte-derived vascular permeability factor (vascular endothelial growth factor) is a potent mitogen for dermal microvasucular endothelial cells. *J. Invest. Dermatol.* 105, 44–50.

Dry, F. W. (1926). The coat of the mouse (*Mus musculus*). *J. Genet.* 16, 287–340.

Enshell-Seijffers, D., Lindon, C., Wu, E., Taketo, M. M., and Morgan, B. A. (2010). β-catenin activity in the dermal papilla of the hair follicle regulates pigment-type switching. *Proc. Natl. Acad. Sci., USA.* 107, 21564–21569.

Fitzpatrick, T. B., Y. Hori, K. Toda, and M. Seiji (1969) Melanin 1969: Some definitions and problems. *Jap. J. Dermatol.* (Ser. B). 79, 278–282.

Furuya, R., Akiu, S., Ideta, R., Naganuma, M., Fukuda, M., and Hirobe, T. (2002). Chages in the proliferative activity of epidermal melanocytes in serum-free primary culture during the development of ultraviolet radiation B-induced pigmented spots in hairless mice. *Pigment Cell Res.* 15, 348–356.

Hachiya, A., Kobayashi, A., Ohuchi, A., Takema, Y., and Imokawa, G. (2001). The paracrine role of stem cell factor/c-kit signaling in the activation of human melanocytes in ultraviolet-B-induced pigmentation. *J. Invest. Dermatol.* 116, 578–586.

Halaban, R., Langdon, R., Birchall, N., Cuono, C., Baird, A., Scott, G., Moellmann, G., and McGuire, J. (1988). Basic fibroblast growth factor from human keratinocytes is a natural mitogen for melanocytes. *J. Cell Biol.* 107, 1611–1619.

Hara, M., Yaar, M., and Gilchrest, B.A. (1995). Endothelin-1 of keratinocyte origin is a mediator of melanocyte dendricity. *J. Invest. Dermatol.* 105, 744-748.

Hearing, V.J. (2000). The melanosome: the perfect model for cellular response to the environment. *Pigment Cell Res.* 13 (Suppl. 8), 23–34.

Hirobe T. (1983). Proliferation of epidermal melanocytes during the healing of skin wounds in newborn mice. *J. Exp. Zool.* 227, 423–431.

Hirobe, T. (1984) Histochemical survey of the distribution of the epidermal melanoblasts and melanocytes in the mouse during fetal and postnatal periods. *Anat. Rec.* 208, 589–594.

Hirobe, T. (1988a) Developmental changes of the proliferative response of mouse epidermal melanocytes to skin wounding. *Development* 102, 567–574.

Hirobe, T. (1988b). Genetic factors controlling the proliferative activity of mouse epidermal melanocytes during the healing of skin wounds. *Genetics* 120, 551–558.

Hirobe, T. (1992a) Control of melanocyte proliferation and differentiation in the mouse epidermis. *Pigment Cell Res.* 5, 1–11.

Hirobe, T. (1992b) Basic fibroblast growth factor stimulates the sustained proliferation of mouse epidermal melanoblasts in serum-free medium in the presence of dibutyryl cyclic AMP and keratinocytes. *Development* 114, 435–445.

Hirobe, T. (1992c). Melanocyte stimulating hormone induces the differentiation of mouse epidermal melanocytes in serum-free culture. *J. Cell. Physiol.* 152, 337–345.

Hirobe, T. (1994). Keratiocytes are involved in regulating the developmental changes in the proliferative activity of mouse epidermal melanoblasts in serum-free culture. *Dev. Biol.* 161, 59–69.

Hirobe, T. (1995). Structure and function of melanocytes: microscopic morphology and cell biology of mouse melanocytes in the epidermis and hair follicle. *Histol. Histopathol.* 10, 223–237.

Hirobe, T. (2001). Endothelins are involved in regulating the proliferation and differentiation of mouse epidermal melanocytes in serum-free primary culture. *J. Invest. Dermatol. Symp. Proc.* 6, 25–31.

Hirobe T. (2002). Role of leukemia inhibitory factor in the regulation of the proliferation and differentiation of neonatal mouse epidermal melanocytes in culture. *J. Cell. Physiol.* 192, 315–326.

Hirobe, T. (2005). Role of keratinocyte-derived factors involved in regulating the proliferation and differentiation of mammalian epidermal melanocytes. *Pigment Cell Res.* 18, 2–12.

Hirobe, T. (2009). Ferrous ferric chloride stimulates the proliferation of human skin keratinocytes, melanocytes, and fibroblasts in culture. *J. Health Sci.* 55, 447–455.

Hirobe, T. (2011a). How are proliferation and differentiation of melanocytes regulated? Pigment Cell Melanoma Res. 24, 462–478.

Hirobe T. (2011b). Stimulation of the proliferation and differentiation of skin cells by ferrous ferric chloride from a distance. *Biol. Pharm. Bull.* 34, 987–995.

Hirobe, T. (2012). Iron and skin health: iron stimulates skin function. in : V. R. Preedy (Ed), Handbook of Diet, Nutrition and the Skin, Wageningen, Wageningen Academic Publishers, the Netherland, 196–214.

Hirobe, T., and Abe, H. (1999). Genetic and epigenetic control of the proliferation and differentiation of mouse epidermal melanocytes in culture. *Pigment Cell Res.* 12, 147–163.

Hirobe, T., and Abe, H. (2000). $ACTH_{4\text{-}12}$ is the minimal message sequence required to induce the differentiation of mouse epidermal melanocytes in serum-free primary culture. *J. Exp. Zool.* 286, 632–640.

Hirobe T., Osawa M., and Nishikawa S-I. (2003) Steel factor controls the proliferation and differentiation of neonatal mouse epidermal melanocytes in culture. *Pigment Cell Res.* 16, 644–655.

Hirobe T., Osawa M., and Nishikawa S-I. (2004a) Hepatocyte growth factor controls the proliferation of cultured epidermal melanoblasts and melanocytes from newborn mice. *Pigment Cell Res.* 17, 51–61.

Hirobe T., Furuya R., Hara E., Horii I., Tsunenaga M., and Ifuku O. (2004b). Granulocyte-macrophage colony-stimulating factor controls the proliferation and differentiation of mouse epidermal melanocytes from pigmented spots induced by ultraviolet radiation B. *Pigment Cell Res.* 17, 230-240.

Hirobe T., Furuya R., Ifuku O., Osawa M., and Nishikawa S-I. (2004c). Granulocyte-macrophage colony-stimulating factor is a keratinocyte-derived factor involved in regulating the proliferation and differentiation of neonatal mouse epidermal melanocytes in culture. *Exp. Cell Res.* 297, 593–606.

Hirobe, T., Takeuchi, S., and Hotta, E. (2004d). The melanocortin receptor-1 gene but not the proopiomelanocortin gene is expressed in melanoblasts and contributes their differentiation in the mouse skin. *Pigment Cell Res.* 17, 627–635.

Hirobe, T., Shinpo, T., Higuchi, K., and Sano, T. (2010). Life cycle of human melanocytes is regulated by endothelin-1 and stem cell factor in synergy with cyclic AMP and basic fibroblast growth factor. *J. Dermatol. Sci.* 57, 123–131.

Horsley, V., O'Carroll, D., Tooze, R., Ohinata, Y., Saitou, M., Obukhanych, T., Nussenzweig, M., Tarakhovsky, A., and Fuchs, E. (2006). Blimp 1 defines a progenitor population that governs cellular input to the sebaceous gland. *Cell* 126, 597–609.

Imokawa, G. (2004). Autocrine and paracrine regulation of melanocytes in human skin and in pigmentary disorders. *Pigment Cell Res.* 17, 96–110.

Imokawa, G., Yada, Y., and Miyagishi, M. (1992). Endothelins secreted from human keratinocytes are intrinsic mitogens for human melanocytes. *J. Biol. Chem.* 267, 24675–24680.

Imokawa, G., Yada, Y., Kimura, M., and Morisaki. N. (1996). Granulocyte/macrophage colony-stimulating factor is an intrinsic keratinocyte-derived growth factor for human melanocytes in UVA-induced melanosis. *Biochem. J.* 313, 625–631.

Imokawa, G., Kobayashi, T., Miyagishi, M., Higashi, K., and Yada, Y. (1997). The role of endothelin-1 in epidermal hyperpigmentation and signaling mechanisms of mitogenesis and melanogenesis. *Pigment Cell Res.* 10, 218–228.

Inomata, K., Aoto, T., Binh, N.T., Okamoto, N., Tanimura, S., Wakayama, T., Iseki, S., Hara, E., Masunaga, T., Shimizu, H., and Nishimura, E. K. (2009). Genotoxic stress abrogates renewal of melanocyte stem cells by triggering their differentiation. *Cell* 131, 1088–1099.

Iriyama, S., Ono, T., Aoki, H., and Amano, S. (2011). Hyperpigmentation in human solar lentigo is promted by heparanase-induced loss of heparan sulfate chains at the dermal-epidermal junction. *J. Dermatol. Sci.* 64, 223–228.

Ito, S. (2003). A chemist's view of melanogenesis. *Pigment Cell Res.* 16, 230–236.

Kobayashi, K., Rochat, A., and Barrandon Y. (1993). Segregation of keratinocyte colony-forming cells in the bulge of rat vibrissa. *Proc. Natl. Acad. Sci., USA* 90, 7391–7395.

Kovacs, D., Cardinali, G., Aspite, N., Cota, C., Luzi, F., Bellei, B., Briganti, S., Amantea, A., Torrisi, M. R., and Picardo, M. (2010). Role of fibroblast-derived growth factors in regulating hyperpigmentation of solar lentigo. *Br. J. Dermatol.* 163, 1020–1027.

Kunisada, T., Yoshida, H., Yamazaki, H., Miyamoto, A., Hemmi, H., Nishimura, E., Shultz, L.D., Nishikawa, S.-I., and Hayashi. S.-I. (1998). Transgene expression of steel factor in the basal layer of epidermis promotes survival, proliferation, differentiation and migration of melanocyte precursors. *Development,* 125, 2915–2923.

Kunisada, T., Yamazaki, H., Hirobe, T., Kamei, S., Omoteno, M., Tagaya, H., Hemmi, H., Koshimizu, U., Nakamura, T., and Hayashi, S.-I. (2000). Keratinocyte expression of transgenic hepatocyte growth factor affects melanocyte development, leading to dermal melanocytosis. *Mech. Dev.* 94, 67–78.

Levy, V., Lindon, C., Zheng, Y., Harfe, B.D., and Morgan, B. A. (2007). Epidermal stem cells arise from the hair follicle after wounding. *FASEB J.* 21, 1358–1366.

Li, W., Fan, J., Chen, M., Guan, S., Sawcer, D., Bokoch, G. M., and Woodley, D. T. (2004). Mechanism of human dermal fibroblast migration driven by type I collagen and platelet-derived growth factor BB. *Mol. Biol. Cell* 15, 294–309.

Lyle, S., Christofidou-Solomidou, M., Liu, Y., Elder, D. E., Albelda, S., and Cotsarelis, G. (1998). The C8/144B monoclonal antibody recognizes cytokeratin 15 and defines the location of human hair follicle stem cells. *J. Cell Sci.* 111, 3179–3188.

Maas-Szabowski, N., Shimotoyodome, A., and Fusenig, N.E. (1999). Keratinocyte growth regulation in fibroblast cocultures via a double paracrine mechanism. *J. Cell Sci.* 112, 1843–1853.

Marchese, C., Messina, A., Faggioni, A., Torrisi, M. R., Frati, L., Ron, D., Rubin, J., and Aaronson, A. (1990). Human keratinocyte growth factor activity on proliferation and differentiation of human keratinocytes: differentiation response distinguishes KGF from EGF family. *J. Cell. Physiol.* 144, 326–332.

Mildner, M., Mlitz, V., Gruber, F., Wojta, J., and Tschachler, E. (2007). Hepatocyte growth factor establishes autocrine and paracrine feedback loops for the protection of skin cells after UV irradiation. *J. Invest. Dermatol.* 127, 2637–2644.

Morris, R. J. (2000). Keratinocyte stem cells: targets for cutaneous carcinogens. J. Clin. Invest. 106, 3–8.

Nishikawa-Torikai, S., Osawa, M., and Nishikawa, S. (2011). Functional characterization of melanocyte stem cells in hair follicles. *J. Invest. Dermatol.* 131, 2358–2367.

Nishimura, E. K., Jordan, S. A., Oshima, H., Yoshida, H., Osawa, M., Moriyama, M., Jackson, I. I., Barrandon, Y., Miyachi, Y., and Nishikawa, S.-I. (2002). Dominant role of the niche in melanocyte stem-cell fate determination. *Nature* 416, 854–860.

Nishimura, E. K., Granter, S. R., and Fisher, D. E. (2005). Mechanisms of hair graying: incomplete melanocyte stem cell maintenance in the niche. *Science* 307, 720–724.

Nishimura, E. K., Suzuki, M., Igras, V., Du, J., Lonning, S., Miyachi, Y., Roes, J., Beerman, F., and Fisher, D. E. (2010). Key roles for transforming growth factor beta in melanocyte stem cell maintenance. *Cell Stem Cell* 6, 130–140.

Osawa, M., Egawa, G., Mak, S.-S., Moriyama, M., Freter, R., Yonetani, S., Beerman, F., and Nishikawa, S.-I. (2005). Molecular characterization of melanocyte stem cells in their niche. *Development* 132, 5589–5599.

Pankonin, M. S., Gallagher, J. T., and Loeb, J. A. (2005). Specific structural features of heparan sulfate sulfate proteoglycans potentiate neuregulin-1 signaling. *J. Biol. Chem.* 280, 383–388.

Patel, V. N., Knox, S. M., Likar, K. M., Lathrop, C. A., Hossain, R., Eftekhari, S., Whitelock, J. M., Elkin, M., Vlodavsky, I., and Hoffman, M. P. (2007). Heparanase cleavage of perlecan heparan sulfate modulates FGF10 activity during ex vivo submandibular gland branching morphogenesis. *Development* 134, 4177–4186.

Paus, R., Maurer, M., Slominski, A., and Czametzki, B. M. (1994). Mast cell involvement in murine hair growth. *Dev. Biol.* 163, 230–240.

Peters, E. M. J., Tobin, D. J., Botchkareva, N., Maurer, M., and Paus, R. (2002). Migration of melanoblasts into the developing murine hair follicle is accompanied by transient c-kit expression. *J. Histochem. Cytochem.* 50, 751–766.

Pincelli, C., and Marconi, A. (2010). Keratinocyte stem cells: friends and foes. *J. Cell. Physiol.* 225, 310–315.

Quevedo, W. C., Jr., and Smith, J. A. (1963). Studies on radiation-induced tanning of skin. *Ann. N. Y. Acad. Sci.* 100, 364–388.

Quevedo, W. C., Jr., and Fleischmann, R. D. (1980). Developmental biology of mammalian melanocytes. *J. Invest. Dermatol.* 75, 116–120.

Rheinwald, J. G, and Green, H. (1977). Epidermal growth factor and the multiplication of cultured human epidermal keratinocytes. *Nature* 265, 421–424.

Schauer, E., Trautinger, F., Koeck, A., Schwarz, A., Bhardwai, R., Simon, M., Ansel, J. C., Schwarz, T., and Luger, T. A. (1994). Proopiomelanocortin-derived peptides are synthesized and released by human keratinocytes. *J. Clin. Invest.* 93, 2258–2262.

Scott, G., Leopardi, S., Printup, S., Malhi, N., Seiberg, M., and LaPoint, R. (2004). Proteinase-activated receptor-2 stimulates prostaglandin production in keratinocytes: analysis of prostaglandin receptors on human melanocytes and effects of PGE_2 and PGF_{2a} on melanocyte dendricity. J. Invest. *Dermatol.* 122, 1214–1224.

Sebollela, A., Cagliari, T. C., Limaverde, G. S. C. S., Chapeaurouge, A., Sorgine, M. H. F., Coelho-Sampaio, T., Ramos, C. H. I., and Ferreira, S. T. (2005). Heparin-binding sites in granulocyte-macrophage colony-stimulating factor. *J. Biol. Chem.* 280, 31949–31956.

Seiji, M., Shimao, K., Birbeck, M. S. C., and Fitzpatrick, T. B. (1963). Subcellular localization of melanin biosynthesis. *Ann. N. Y. Acad. Sci.* 100, 497–533.

Silver, A. F., Chase, H. B., and Potten, C. S. (1969). Melanocyte precursor cells in the hair follicle germ during the dormant stage (telogen). *Experientia* 25, 299–301.

Slominski, A., and Paus, R. (1993). Melanogenesis is coupled to murine anagen: toward new concepts for the role of melanocytes and the regulation of melanogenesis in hair growth. *J. Invest. Dermatol.* 101, 90S–97S.

Slominski A., Paus R., and Costantino R. (1991). Differential expression and activity of melanogenesis-related proteins during induced hair growth in mice. *J. Invest. Dermatol.* 96, 172–179.

Slominski, A., Paus, R., Plonka, P., Chakraborty, A., Maurer, M., Pruski, D., and Lukiewicz, S. (1994). Melanogenesis during the anagen-catagen-telogen transformation of the murine hair cycle. *J. Invest. Dermatol.* 102, 862–869.

Slominski, A., Szczesniewski, A. and Wortsman, J. (2000). Liquid chromatography-mass spectrometry detection of corticotropin-releasing hormone and proopiomelanocortin-derived peptides in human skin. *J. Clin. Endocrinol. Metab.* 85, 3582–3588.

Sugiyama, S. (1979). Mode of redifferentiation and melanogenesis of melanocytes in mouse hair follicles. *J. Ultrastruc. Res.* 67, 40–54.

Szabo, G. (1967). The regional anatomy of the human integument with special reference to the distribution of hair follicles, sweat glands and melanocytes. Phil. Trans. R. Soc. London (Biol.) 252, 447–485.

Tada, A., Suzuki, I., Im, S., Davis, M. B., Cornelius, J., Babcock, G., Nordlund, J. J., and Abdel-Malek Z. (1998). Endothelin-1 is a paracrine growth factor that modulates melanogenesis of human melanocytes and participates in their responses to ultraviolet radiation. *Cell Growth Differ.* 9, 575–584.

Tamm, I., Kikuchi, T., and Zychlinsky, A. (1991). Acidic and basic fibroblast growth factors are survival factors with distinctive activity in quiescent BALB/c 3T3 murine fibroblasts. *Proc. Natl. Acad. Sci., USA* 88, 3372–3376.

Takeuchi, T. (1985). Genes controlling intercellular communication during melanocyte differentiation in the mouse. *Zool. Sci.* 2, 823–831.

Thody, A. J., Ridley, K., Penny, R. J., Chalmers, R., Fisher, C., and Shuster, S. (1983). MSH peptides are present in mammalian skin. *Peptides* 4, 813–816.

Tobin, D. J. (2006). Biochemistry of human skin—our brain on the outside. *Chem. Soc. Rev.* 35, 52–67.

Virador, V. M., Muller, J., Wu, X., Abdel-Malek, Z.A., Yu, Z-X., Ferrans, V. J., Kobayashi, N., Wakamatsu, K., Ito, S., Hammer, J. A., and Hearing, V. J. (2001). Influence of α-melanocyte-stimulating hormone and of ultraviolet radiation on the transfer of melanosomes to keratinocytes. *FASEB J.* 15, U135–U161.

Wakamatsu, K., Graham, A., Cook, D., and Thody, A. J. (1997). Characterisation of ACTH peptides in human skin and their activation of the melanocortin-1 receptor. *Pigment Cell Res.* 10, 288–297.

Whitelock, J. M., and Iozzo, R. V. (2005). Heparan sulfate: a complex polymer charged with biological activity. *Chem. Rev.* 105, 2745–2764.

Yaar, M., Grossman, K., Eller, M., and Gilchrest, B. A. (1991). Evidence for nerve growth factor-mediated paracrine effects in human epidermis. *J. Cell Biol.* 115, 821–828.

Yada, Y., Higuchi, K., and Imokawa, G. (1991). Effects of endothelins on signal transduction and proliferation in human melanocytes. *J. Biol. Chem.* 266, 18352–18357.

Yohn, J. J., Morelli, J. G., Walchak, S. J., Rundell, K. B., Norris, D. A., and Zamora, M. R. (1993). Cultured human keratinocytes synthesize and secrete endothelin-1. *J. Invest. Dermatol.* 100, 23–26.

Yu, H., Kumar, S. M., Kossenkov, A. V, Showe, L., and Xu, X. (2010). Stem cells with neural crest characteristics derived from the bulge region of cultured human hair follicles. *J. Invest. Dermatol.* 130, 1227–1236.

In: Keratinocytes
Editor: Elia Ranzato

ISBN: 978-1-62618-798-6
© 2013 Nova Science Publishers, Inc.

Chapter 3

Keratinocyte Differentiation: Focus on MAPK Signal Transduction

Tanja Popp[*] *and Christian Ries*
Institute for Cardiovascular Prevention
Ludwig-Maximilians-University of Munich (LMU)

Abstract

The epidermis is a multilayer epithelium composed of keratinocytes representing the first line of defense in the skin. As a consequence to permanent environmental stress, the epidermis is subjected to a constant renewal. Immature keratinocytes present in the basal layer of the epidermis, so called interfollicular stem cells, are able to proliferate and migrate to the outermost granular layer of the skin where they replace fully differentiated keratinocytes, so called corneocytes.

The differentiation of keratinocytes *in vivo* is largely controlled by a calcium (Ca^{2+}) gradient present in the epidermis with low levels of Ca^{2+} in the basal cell layer and high levels of Ca^{2+} in the outer cell layer. The progression of keratinocyte maturation is tightly regulated and characterized by the expression of specific differentiation marker proteins in these cells.

---

[*] E-mail address: tanja.popp@med.uni-muenchen.de.

The mitogen-activated protein kinase (MAPK) is essentially involved in the control of the complex life cycle of keratinocytes. Dysregulation in the molecular mechanisms of keratinocyte maturation contributes to the pathophysiology of various human skin diseases such as psoriasis.

Introduction

The skin is the largest organ in humans covering the entire surface of the body (Goldsmith, 1990). Above all its multiple tasks, the skin provides a protective border between internal and external environments. Thereby it regulates body temperature and the amount of water released from the body. The skin protects from invading pathogens such as bacteria, viruses and chemical substances but also allows transmission of signals from external stimuli (Blanpain et al., 2009). The various functions of skin are clearly reflected by its complex structure and composition. Principally, the skin consists of the epidermis and the underlying dermis that are separated by a basement membrane. The epidermis is composed of the outermost layers of skin cells, the keratinocytes. Since the epidermis is completely renewed every 40-56 days, proliferation, differentiation and migration are critical processes in keratinocytes (Halprin, 1972).

The ability of keratinocytes for directed migration is of crucial importance not only in the normal regeneration of the epidermis but also in injury and inflammation of skin tissues. A functional cutaneous wound healing process is essential to ensure a rapid and efficient restoration of the protective epithelium. This repair procedure in the dermis and epidermis can be subdivided into a series of events involving numerous tightly controlled cellular and molecular mechanisms (Martin, 1997). At sites of injury, (pro-)inflammatory cytokines and chemokines promote the recruitment of immune cells which in turn secrete various bioactive mediators that attract fibroblasts from dermis into the wound bed. The fibroblasts then build up a provisional structure of extracellular matrix which provides a scaffold for migrating cells. Concurrently, keratinocytes start to proliferate and migrate from the wound margins along the new collagen network to support the re-epithelialization process. In late phases of wound healing, fibroblasts reorganize the wound matrix by contraction and remodeling of the extracellular matrix involving the activity of various proteases including the matrix metalloproteinases and their inhibitors (Vu et al., 2000).

At all stages of skin regeneration and wound healing, the growth, maturation and migration of keratinocytes must be tightly controlled to make sure that these processes proceed in a physiological manner. The present chapter provides insight into the current knowledge about the molecular mechanisms governing the differentiation of interfollicular stem cells into fully mature keratinocytes with a focus on the role of MAPK signaling in the regulation of this process.

Skin Organisation

Epidermis

The epidermis represents a multi-layer epithelium (Figure 1). The outermost cell layer consists of dead keratinocytes. These so-called corneocytes are stably interconnected via desmosomes and thereby provide an effective barrier against chemical and mechanical attacks from the outside. The underlying stratum granulosum is build up from keratinocytes that contain characteristic granules such as the lamellar granules and keratohyalin granules giving this cell layer its typical granulated look (Candi et al., 2005).

The lamellar granules harbor a variety of lipids. Upon exocytosis from the cells, these lipids contribute to a hydrophobic barrier that prevents diffusion of water and invasion of potentially harming substances from the outside into the skin (Jungersted et al., 2008). The keratohyalin granules contain specific proteins including profillagrin and loricrin which after secretion from the keratinocytes can bundle intermediate filaments and thus contribute to the strength of the epidermis. In biochemical cell analysis, profillagrin and loricrin serve as late differentiation markers to identify mature keratinocytes (Fuchs, 2007).

Due to permanent stress, the epidermis is subjected to a constant renewal process that is based on the activity of immature keratinocytes. These so-called interfollicular stem cells–reside in the stratum basale. They are mitotically active and characteristically express proliferating cell nuclear antigen (PCNA) which is important for DNA replication in the S phase (Tsuji *et al.*, 1998). Their nascent daughter cells are called transient amplifying cells. They undergo a differentiation process while traveling through the stratum spinosum and stratum granulosum to the outermost layer of the skin. Eventually, the fully mature keratinocytes are subject to a desquamation process resulting in their shedding as dead cells from the skin surface (Kypriotou et al., 2012) (Figure 1).

Common skin diseases such as psoriasis and atopic dermatitis are characterized by defects of the above described keratinocyte ontogeny (Ambler et al., 2009; Jensen et al., 2004).

Dermis

The dermis is the skin layer located directly under the epidermis. In contrast to the epidermis which primarily consists of a specific cell type albeit at different stages of differentiation, the dermis has a more heterogeneous composition. It harbors skin´s appendages, nerves, vessels and various dermal cells including fibroblasts, macrophages and adipocytes. These cells provide the proper environment for hair follicles which are fixed by the erector pili muscles attached to each follicle. Within these structures a so called bulge region harbors specified stem cells which contribute to normal hair life cycle and participate in wound repair (Levy et al., 2007). All dermal structures are embedded in a characteristic extracellular matrix mainly provided by fibroblasts which are the predominant cell type in this layer. Collagen type I represents the most abundant protein in the dermis. It forms three dimensional networks together with collagen type III and V thus providing the elasticity, strength and overall structural integrity of the skin (Myllyharju et al., 2004; Schultz et al., 2009).

Basement Membrane

Both fibroblasts and keratinocytes secrete proteins required for the assembly of basement membranes which connect the epidermis to the dermis and thereby ultimately allow the formation of a tight epithelium. Basement membranes have a unique net-like structure containing several different proteins, most abundantly laminin, collagen type IV, collagen type VII, nidogen and perlecan (Marinkovich et al., 1993). Hemidesmosomes are small patches present on the inner basal surface of keratinocytes anchoring the stratum basale to the basement membrane. A typical hemidesmosome consists of cytosolic keratin bonded to a transmembrane adhesion molecule including $\alpha6\beta4$ integrin which can bind to an adhesive protein in the basement membrane such as laminin-332 (Litjens et al., 2006). **In addition, $\beta1$ integrins** are found in stem cells of the basal layer predominantly at sites of focal adhesion to the basement membrane. Interestingly, the β1 integrins are also essential in the regulation of keratinocyte life cycle where they serve as a negative stimulus for differentiation (Levy et al., 2000; Margadant et al., 2010). The different patterns and levels of integrin expression in keratinocytes are exploited to biochemically distinguish interfollicular stem cells from more differentiated progeny cells.

Keratinocyte Differentiation

Epidermal Stem Cells and their Progenies

The four major sublayers of the epidermis consist of keratinocytes that differ by shape and expression of typical marker proteins indicating their distinct stages of cellular differentiation. The cells in the lowermost so-called basal layer are mitotically active and represent the interfollicular stem cell population (Watt et al., 2009). A certain portion of their daughter cells can migrate to the outermost layer of the skin where they undergo a maturation process of terminal differentiation and ultimately form the surface of the epidermis. A key issue in this context is the homeostatic control of interfollicular epidermal stem cell proliferation and differentiation in the stratum basale. Formerly, it was assumed that interfollicular stem cells represent a homogenous cell population that mainly divide in a symmetrical manner to maintain the stem cell pool but randomly differentiate into progeny cells destined to move to the outward layers (Epstein et al., 1965; Marques-Pereira et al., 2012). At present, the most widely accepted model for skin stem cell homeostasis is based on asymmetrical division of interfollicular stem cells (Fuchs et al., 2012). This results in the generation of both replicate stem cells and more differentiated progenies, so-called transit amplifying cells. They are organized into columns named epidermal proliferative units and destined to undergo terminal differentiation after several rounds of cell division (Lechler et al., 2005; Potten, 1974). Although interfollicular stem cells have a pronounced capacity for reproduction, events of cell division occurs only infrequently in the basal layer of the epidermis. This is reflected by the low number of cells positive for expression of the proliferation marker Ki67 (Usui et al., 2008). The additional existence of transit amplifying cells and their progenies in the stratum basale could explain the delivery of sufficient amounts of differentiated keratinocytes required in the skin. A recent review article provides intense discussion on the current concepts and molecular mechanisms of epidermal maintenance by skin stem cells (Clayton et al., 2007; Doupe et al., 2012; Mascre et al., 2012).

After leaving the stem cell niche in the stratum basale, migrating keratinocytes encounter increasing concentrations of calcium ions (Ca^{2+}). This Ca^{2+} gradient present in the epidermal cell layers was recognized not only to correlate with the stage of differentiation in the keratinocytes but also regulate this process (Elias et al., 2002) (Figure 1). During maturation, keratinocytes dramatically change in appearance and intracellular gene expression profiles. Morphologically, keratinocytes undergoing terminal differentiation transform from a cylindrical aspect to a flattened phenotype that is accompanied by a reorganization of the cytoskeleton.

Protein Expression in Differentiating Keratinocytes

Keratins are the most abundant components of the intracellular cytoskeleton in epithelial cells. The keratins are composed of type I and type II subunits which build heterodimeric complexes that are packed in oligomers forming intermediate filaments. The intermediate filaments are cross-linked by disulfide bridges which build up a dense mesh of cytoplasmic fibers connected to desmosomes at sites of cell-cell-contact thereby contributing to strength and stability of cells and cell layers (Simpson , 2011). Genetic defects in keratins are implicated in the pathology of diseases such as epidermolysis bullosa and epidermolytic hyperkeratosis (Coulombe et al., 2012; Lane et al., 2004). Both onset and progression of keratinocyte differentiation are characterized by changes in the composition of intermediate filaments. Immature keratinocytes in the stratum basale show an innate production of keratin 5 and keratin 14 (Alam et al., 2011; Fuchs et al., 1980). In contrast, their progeny cells present in the spinous layer switch over to a biosynthesis of keratin 1 and keratin 10 (Figure 1). This alteration in keratin expression is utilized as biochemical marker for the initiation and early progression of keratinocyte differentiation (Porter et al., 2003).

Strikingly, many genes required for the development of keratinocytes into the cornified envelope (CE) are organized in a cluster on chromosome 1, the so-called epidermal differentiation complex (EDC) (Volz et al., 1993). The EDC is composed of three major gene families, namely CE precursor genes, S100 genes and S100 fused genes (Kypriotou et al., 2012).

The most abundant CE precursors are involucrin, loricrin and small proline-rich proteins (Steinert et al., 1998). Involucrin plays an important role during the early steps of CE formation (Eckert et al., 1986). In terminally differentiating keratinocytes it serves as a scaffold protein for ceramide lipids to become covalently attached to the protein envelope on the outer surface of the cells (Marekov et al., 1998). Loricrin represents a major component of the CE (Hohl et al., 1991). After its production in terminally differentiating keratinocytes, loricrin is deposited in the keratohyalin granules and subsequently cross-linked by the action of transglutaminases thereby contributing to the mechanical resistance of the protein net in the CE et al., 2005). Also the group of proline-rich proteins are subjected to transglutaminase activity thus getting interconnected to the other components of the CE (Tesfaigzi et al., 1999).

The S100 family consists of a variety of low molecular weight proteins with a bunch of functions in keratinocytes (Eckert et al., 2004a). All members of the S100 family are dimeric proteins characterized by their Ca^{2+}-binding EF-hand motifs. The interaction of Ca^{2+} with S100 proteins leads to a structural rearrangement of the helices thereby opening a cleft in which the recognition site for their target protein resides (Zimmer et al., 2003). S100 proteins are reported to exhibit various modes of action. Some S100 representatives can inhibit protein phosphorylation by blocking the access of the kinase to its substrate, others regulate Ca^{2+} homeostasis or contribute to cytoskeleton organization (Donato, 2001). Expression analyses have demonstrated that at least 11 members of the S100 family are synthesized in the skin (Eckert et al., 2004a).

S100A10, a member of the S100A subfamily of S100 genes, is a known substrate of transglutaminases in terminally differentiating keratinocytes and found to be covalently integrated in the protein net of the CE (Robinson et al., 1997; Ruse et al., 2001). Furthermore, S100A10 is part of a complex with annexin II and calpactin I. This complex is localized at the plasma membrane where it forms a Ca^{2+} channel that facilitates influx of Ca^{2+} thus contributing to Ca^{2+}-dependent keratinocyte maturation (Broome et al., 2003; Eckert et al., 2004a).

Keratinocytes in the granular cell layer of the epidermis typically produce the keratohyalin granules that contain profillagrin which belongs to the third family of the EDC, the S100 fused genes (Candi et al., 2005). Profillagrin is a 400 kDa, histidine-rich and highly phosphorylated precursor protein. Its processed form named fillagrin is needed for the aggregation of keratins to build up tight bundles of intermediate filaments that are characteristically found in cornified cells (Steinert et al., 1981). The proteolytic degradation of fillagrin in the stratum corneum generates amino acids and further metabolites that serve as natural moisturizing and pH regulatory factors (Kypriotou et al., 2012; Rawlings et al., 1994).

Transglutaminases represent a class of enzymes which form covalent crosslinks between proteins. Especially in surface epithelia such as skin transglutaminases catalyze the formation of a tight protein meshwork important for the barrier function of this organ. In keratinocytes, at later stages of differentiation the Ca2+-dependent activity of transglutaminases is upregulated (Candi et al., 2005). Thus it is no surprise that in the epidermis most of the transglutaminase substrates are located in the CE. Transglutaminases for instance attach involucrin to certain desmosomal proteins that allow cell-to-cell adhesion at the lateral side of the plasma membrane (Eckert et al., 2005). This scaffold is further stabilized by transglutaminase-mediated incorporation of ceramides, reinforcement proteins such as loricrin and by cross-linkage of keratin filaments (Hitomi, 2005). Transglutaminase 1 is the predominant form of transglutaminases in keratinocytes and its expression is initiated in the cells when entering the granular layer (Iizuka et al., 2003; Michel et al., 1992). Mice lacking transglutaminase 1 suffer from a markedly impaired skin barrier and die within a few days as a consequence of dehydration (Matsuki et al., 1998).

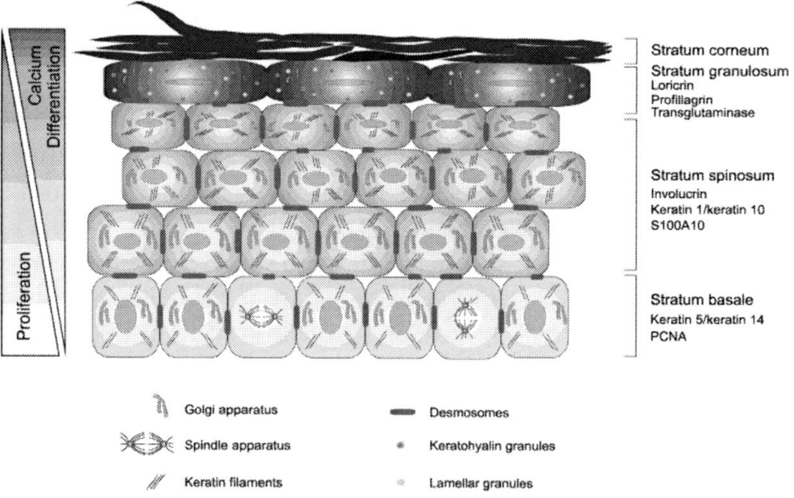

Figure 1. Illustration of the epidermal structure and depiction of characteristic proteins expressed in keratinocytes during terminal differentiation by increasing concentrations of Ca^{2+}.

Outside-in Signaling of Ca^{2+} in Keratinocyte Differentiation

It is well established that increasing concentrations of Ca^{2+} present in the outer layers of the epidermis represent the primary stimulus for keratinocyte maturation *in vivo* (Elias et al., 2002). The current concept of molecular pathways that transform extracellular Ca^{2+} levels into intracellular response includes plasma membrane channels and receptors associated with intracellular signaling cascades that eventually modulate gene expression in differentiating keratinocytes.

The transient receptor potential cation (TRPCs) channels are proteins that form ion channels in the plasma membrane (Ramsey et al., 2006). A recent report demonstrated that TRPC6 channels increase the influx of Ca^{2+} into keratinocytes and thereby specifically promote the maturation of these cells (Muller et al., 2008). In an animal model for psoriasis, low expression of TRPC channels was found to be associated with an impaired Ca^{2+}-induced production of keratin 10 and transglutaminase 1 as well as an enhanced proliferation in keratinocytes (Leuner et al., 2011). In addition to TRPC6, the transient receptor potential vanilloid 6 (TRPV6) channel can form highly selective Ca^{2+} channels essential for keratinocyte maturation (Lehen'kyi et al., 2007). A specific knock-down of TRPV6 resulted in an altered keratinocyte morphology accompanied by reduced expression of differentiation markers and loosened cell-cell contacts. Furthermore, 1,25-dihydroxyvitamin D3 that is known to stimulate keratinocyte maturation (Smith et al., 1986; Su et al., 1994) was shown to enhance TRPV6 biosynthesis in these cells et al., 2007). This finding is in line with the concept of a TRPV6 participation in the differentiation process by increasing the Ca^{2+} influx in keratinocytes. Consistently, TRPV6 knock-out mice show a disturbed Ca^{2+} gradient in the epidermis and exhibit fewer and thinner layers of the stratum corneum (Bianco et al., 2007).

Ion channels such as TRPC6 and TRPV6 directly augment cytoplasmatic Ca^{2+} by increasing the influx into the cell. The discovery of a specific calcium sensing receptor (CaSR) that belongs to the family of G protein-coupled receptors (GPCRs) (Bouschet et al., 2005) introduced the concept of a pathway which non-directly transforms pericellular Ca^{2+} levels into an intracellular response. The knock-out of CaSR in mice provoked a highly disturbed epidermal development (Tu et al., 2012) indicating the importance of this protein in keratinocyte differentiation. The molecular mechanism of CaSR function involves binding of Ca^{2+} to the receptor on the surface of the cells, intracellular recruitment of heterotrimeric $G\alpha_q$ proteins to the inner part of the receptor and subsequent activation of phospholipase C resulting in the release of Ca2+ from intracellular storage organelles (Bouschet et al., 2005). Elevated levels of intracellular Ca^{2+} independent of the producing mechanism then lead to the activation of protein kinase C (PKC) (Breitkreutz et al., 2007) (Figure 2).

A large number of PKC isoforms has been identified to be differently distributed throughout the specific cell layers of the epidermis (Denning, 2004; Fisher et al., 1993). In particular, the three PKC isoforms PKCα, PKCδ and PKCη are highly expressed in keratinocytes where they perform downstream signaling involving MAPK activity (Adhikary et al., 2010; Breitkreutz et al., 2007; Denning et al., 1995).

MAPK Signaling in Keratinocyte Differentiation

The Mapk Signaling Cascades

The MAPK signaling pathways play pivotal roles in the regulation of various essential processes in eukaryotic cells including differentiation, proliferation and migration. The principal mechanism of MAPK signaling is as follows. Cytokines, growth factors and other stimulatory factors bind to their specific receptors at the plasma membrane. Subsequently, the intracellular transmission of signaling occurs by three sequential phosphorylation steps resulting in an activating cascade involving MAP kinase kinase kinases (MAPKKK), downstream MAPKKs (also named MEKs) and finally the MAPKs (Pearson et al., 2001). The most prominent MAPKs are p38, the Jun N-terminal kinase (JNK) and the extracellular signal-regulated kinases 1 and 2 (ERK1/2) giving name to their respective signaling pathway. The phosphorylated MAPKs translocate into the nucleus where they modulate gene expression by phosphorylating target proteins including various transcription factors, which, in turn, control the expression of specific target genes (Pearson et al., 2001). In MAPK pathways the characteristic arrangement of sequentially activating kinases allows the amplification of weak external signals and provides the opportunity for several levels of intervention by feedback regulatory mechanisms (Legewie et al., 2005; Pearson et al., 2001).

The Signaling Pathway of p38

p38 and JNK are frequently categorized as so-called stress-induced kinases because their signaling pathways respond to similar stimuli including UV light and the pro-inflammatory cytokines tumor necrosis factor α and interleukin 1 (Heinrichsdorff et al., 2008; Silvers et al., 2003; Wolfman et al., 2002; Yuan et al., 2003). Several studies on cellular signaling mechanisms involved in tissue inflammation demonstrated that p38 plays a key role in the pathophysiology of this process (Thalhamer et al., 2008; Yong et al., 2009). Moreover, p38 deficiency in mice results in embryonic lethality highlighting the importance of p38 activity in normal physiological development (Allen et al., 2000).

The p38 signal transduction pathway can be induced not only by a variety of extracellular stimuli including physiological and chemical stress but also by elevated levels of extracellular Ca^{2+}. Epidermal keratinocytes are able to sense and respond to changes in the Ca^{2+} concentration by their membrane receptors that pass on the signal via downstream PKCs. Efimova and colleagues identified PKCδ and PKCη, members of the family of novel PKCs, that specifically activate the p38 pathway in keratinocytes (Efimova et al., 2002). Both PKCδ and PKCη strongly enhance the phosphorylation of the p38 target protein ATF-2 without changing the total amount of p38 protein. Although the activation of p38 by the classical PKCα was less prominent when compared to PKCδ and PKCη, the knock-down of PKCα in primary human keratinocytes abrogated the production of involucrin and transglutaminase 1 (Yang et al., 2003). Consistently, overexpression of PKCα in both its normal and active form stimulated the biosynthesis of the differentiation markers keratin 1, loricrin and filaggrin as demonstrated in mouse keratinocytes (Seo et al., 2004). Inversely, p38-deficient keratinocytes showed a tremendous reduction in involucrin gene expression (Dashti et al., 2001a), suggesting importance of p38 activity in keratinocytes differentiation (Figure 2). This was confirmed by own studies demonstrating that the inhibition of the isoforms p38α and p38β prevented the transcription and biosynthesis of keratin 1 and loricrin during Ca^{2+}-induced differentiation in keratinocytes (Popp et al., 2011).

Lipid rafts are micro domains in plasma membranes that organize centers for signaling molecules (Simons et al., 2000). The depletion of cholesterol leads to a prolonged phosphorylation of p38α and p38β accompanied by the upregulation of involucrin. This suggests that the specific composition of lipid rafts modulates the phenotype of early differentiating keratinocytes by regulation of the activity of signaling proteins known to control keratinocyte differentiation (Jans et al., 2004). Furthermore, p38δ was reported to be mandatory for a proper involucrin promoter activity during differentiation of immature keratinocytes stimulated by 12-O-tetradecanoyl-phorbol-13-acetate and okadaic acid (Adhikary et al., 2010; Efimova et al., 2003).

Interestingly, *ex vivo* expansion of epithelial progenitor cells is improved in the presence of p38 inhibitors indicating a role of this MAPK in the switch from cell proliferation to differentiation (Peng et al., 2009). Maturation of other cell types such as myoblasts and osteoclasts was also shown to be regulated by p38 activity (Al-Shanti et al., 2008; Matsumoto et al., 2000) suggesting a more generalized function of p38 in the cellular differentiation processes.

Studies on downstream target genes of p38 activity revealed that p38 is able to phosphorylate heat shock protein 27 (Hsp27) that is essential for proper differentiation of keratinocytes (Jonak et al., 2011; Popp et al., 2011). Both inhibition of p38 and knock-down of Hsp27 resulted in an irregular differentiation with a delayed keratin 10 expression and the absence of flattened keratinocytes in the upper epidermis (Jonak et al., 2011).

The Signaling Pathway of ERK1/2

Several cytokines including epidermal growth factor and transforming growth factor β are known to stimulate the ERK1/2 signal transduction pathway and thereby regulate growth and differentiation of cells (Lee et al., 2004; McKay et al., 2007). This occurs by cytokine binding to their specific receptors which frequently involves activation of integrin molecules on the cell surface. Results from several studies in keratinocytes suggest that ERK1/2 activity is a promoter of proliferation in these cells. This effect seems to be closely associated with the expression of particular integrins involved in the regulation of keratinocyte proliferation and differentiation. High expression levels of β1 integrin in proliferating progenitor keratinocytes inversely correlated with the progress of maturation in these cells (Jones et al., 1993). Contrariwise, abrogation of β1 integrin expression reduced ERK1/2 signaling and prompted keratinocytes to exit the cell cycle (Kuwano et al., 2007; Zhu et al., 1999a). In consistence with these findings, the activation of ERK1/2 signaling via integrins was shown to control the affiliation of keratinocytes to the proliferative compartment in the epidermis (Haase et al., 2001; Hobbs et al., 2004).

Increased proliferation of keratinocytes is also a prerequisite for proper wound healing. ERK1/2 signaling was demonstrated to essentially control the cell division and migration of keratinocytes and thereby enables full re-epithelialization of the skin (Shibata et al., 2012). A disturbance of the tight miR-124- and miR-214-dependent inhibitory regulation of ERK1/2 was shown to contribute to a hyperproliferation and malignant transformation in cutaneous squamous cell carcinoma (Yamane et al., 2012).

In contradiction to the evidences indicating that ERK1/2 activity stimulates proliferation and inhibits maturation in keratinocytes, there are *in vitro* reports demonstrating that Ca^{2+}-mediated activation of the ERK1/2 signaling cascade in these cells resulted in an augmented expression of involucrin and p21/Cip1, a cyclin-dependent kinase inhibitor that contributes to cell cycle arrest (Roper et al., 2001; Schmidt et al., 2000). Findings obtained from several *in vivo* studies have greatly enhanced the understanding of ERK1/2´s role in keratinocyte differentiation. In a transgenic mouse model, the expression of a defective Ras gene was set either under the promoter control of keratin 14 that is characteristically expressed in the basal layer of the epidermis or under the promoter control of keratin 1 which is synthesized in the stratum spinosum. Interestingly, mice with a Ras-mediated defect in ERK1/2 signaling of cells present in the basal layer displayed a very thin and shiny skin, and finally died within a few days. In contrast, mice carrying the Ras loss-of-function mutation in cells of the layer spinosum were phenotypically normal (Dajee et al., 2002). These results indicate a layer-specific prevalence of the ERK1/2 pathway activity in the skin with a pivotal role of ERK1/2 for proper keratinocyte function in the proliferative zone of the epidermis. In agreement with these findings, the constitutive upregulation of active Ras in cells of the basal layer caused a massive hyperplasia accompanied by the absence of differentiated cell layers. Further detailed analysis of the immature keratinocytes that overexpress Ras confirmed that ERK1/2 activity promotes proliferation and simultaneously blocks differentiation in these cells (Dajee et al., 2002) (Figure 2).

The Signaling Pathway of JNK

In contrast to p38 and ERK1/2 little is known about the role of JNK and its potential control by PKC in keratinocyte differentiation. It has been reported that the abrogation of JNK activity by use of specific inhibitors or by a gene knock-out promoted differentiation in keratinocytes indicating a negative regulatory function of JNK activity in this process (Dashti et al., 2001b; Gazel et al., 2006) (Figure 2). Previous studies on the role of PKC in this process did not provide evidence for an influence of PKC activity on JNK target gene expression (Efimova et al., 2002). However, recent data collected in a PKCη-null mouse model demonstrated that JNK activity was enhanced in the PKCη-deficient keratinocytes, suggesting an inhibitory influence of PKCη on JNK-mediated effects in these cells (Hara et al., 2011). Further detailed studies are required to understand the role of JNK in the control of keratinocyte maturation.

Crosstalk between MAPK Signaling Pathways

In addition to numerous studies reporting on the role of the particular MAPK pathways in cell function, several recent publications provide evidence for reciprocal interference of the p38, ERK1/2 and JNK signaling activities.

p38 is reported to exhibit negative influence on ERK1/2 activity as demonstrated in various cell types (Eckert et al., 2004b; Efimova et al., 2003; Zhang et al., 2001) (Figure 2). In corneal epithelial cells p38 activates specific phosphatases that initiate dephosphorylation of ERK1/2 and thus downregulate its activity (Wang et al., 2006). Studies in HeLa cells showed a molecular interaction of activated p38 and ERK1/2 (Zhang et al., 2001). Obviously, the complex formation of p38 and ERK1/2 results in the activation of the former and a simultaneous inactivation of the latter as demonstrated in keratinocytes (Efimova et al., 2003). These results are supported by own data revealing that inhibition of activated p38 in human primary keratinocytes not only reduced phosphorylation of the p38 substrate Hsp27 but also augmented the generation of phosphorylated ERK1/2. Conversely, blockage of ERK1/2 signaling in the keratinocytes caused an enhancement of p38 activity as indicated by increased phosphorylation of Hsp27 (Popp et al., 2011). These findings suggest the existence of a bidirectional negative crosstalk mechanism between p38 and ERK1/2 pathway activities in keratinocytes. Further studies on the roles of p38 and ERK1/2 in keratinocyte differentiation elucidated a synchronized mechanism by which Ca^{2+} initiates a stable activation of p38 in the cells that is counteracted by only short-time upregulation of ERK1/2 which is further attenuated through sustained p38 activity during the negative crosstalk between both kinases (Breitkreutz et al., 2007; Dashti et al., 2001b; Denning, 2010; Efimova et al., 2003; Popp et al., 2011; Schmidt et al., 2000) (Figure 2).

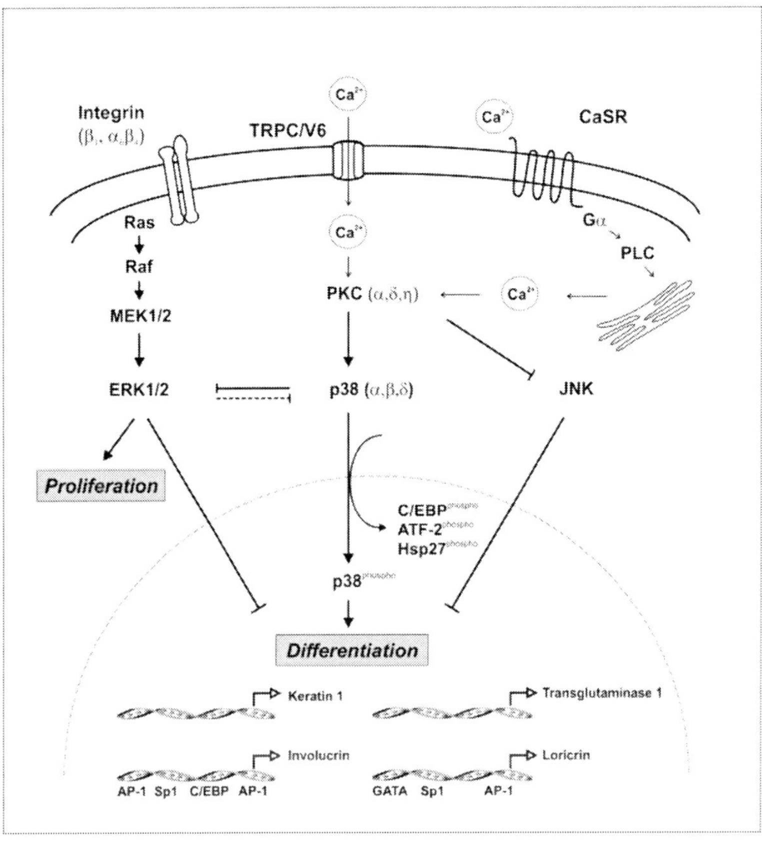

Figure 2. Cell surface receptors, intracellular signaling pathways and target genes important in keratinocyte differentiation and proliferation.

Besides that, liver cells deficient in p38 protein had a significantly increased JNK activity (Heinrichsdorff et al., 2008). Additionally, inhibitor studies in dermal fibroblasts showed that blockage of p38 enhanced and prolonged the activity of JNK (Park et al., 2011). Studies on cell function in hepatocytes and fibroblasts revealed that p38 activity impedes the JNK-dependent proliferation in these cells (Hui et al., 2007; Kayahara et al., 2005; Nakagawa et al., 2012). Fey and colleagues recently proposed a concept that explains p38 inhibition of JNK by an indirect inhibitory mechanism involving ERK1/2 activation as an intermediate step (Fey et al., 2012). These findings clearly suggest that p38 activity negatively influences the function of JNK. If this holds also true for keratinocytes has to be shown in further studies.

MAPK-Mediated Regulation of Gene Transcription

Phosphorylated MAPKs translocate into the nucleus where they phosphorylate target proteins including various transcription factors which, in turn, control the expression of specific genes associated with cell differentiation. Gazel and colleagues reported on a detailed analysis of several transcription factors in terms of their activation by p38, JNK and ERK1/2 and the corresponding binding sites in the promoter of the regulated genes (Gazel et al., 2006; Gazel et al., 2008). p38 was demonstrated to phosphorylate several transcription factors required for the expression of marker proteins that are characteristically synthesized at the different stages of keratinocyte maturation such as involucrin. The promoter of involucrin contains binding sites for CCAAT/enhancer-binding protein (C/EBP) in the proximal and for Sp1 in the distal regulatory regions (Eckert et al., 1986) (Figure 2). In addition, one of the most common transcription factors, activator protein 1 (AP-1), can attach to both the proximal and distal region of the involucrin promoter (Eckert et al., 2004b). The nuclear accumulation of transcription factors which bind to AP-1 sites including c-fos, junB and junD were observed to correlate with morphological changes in differentiating keratinocytes (Adhikary et al., 2010). Consistently, suppression of AP-1 function caused a reduced promoter activity and consequently decreased amounts of involucrin mRNA and protein in cultured human keratinocytes and in the mouse epidermis (Han et al., 2012; Rorke et al., 2010). Similar to involucrin, the gene expression of loricrin is also largely controlled via AP-1 binding sites present in its promoter. More precisely, transactivation of the loricrin gene is mediated by c-fos and junD whereas junB downregulates the promoter activity (Rossi et al., 1998). Besides an additional Sp1 binding site, the loricrin gene harbors a GATA binding motif in the proximal region of its promoter (Figure 2). Results obtained from cotransfection experiments indicate that GATA-3 activates the transcription of loricrin in a cooperative manner together with c-fos and Sp1 (Kawachi et al., 2012) (Figure 2).

Another transcription factor that is activated by the PKC/p38 cascade is C/EBPα. It was identified to mediate differentiation-dependent gene expression in keratinocytes (Efimova et al., 2002). The overexpression of C/EBPα caused a massive increase of involucrin promoter activity. In mouse keratinocytes of the stratum basale the expression of both C/EBPα and its isoform C/EBPβ were strongly upregulated during their differentiation and migration to the outer epidermal cell layers (Lopez et al., 2009). Previous work demonstrated that augmentation of C/EBPβ in keratinocytes induced the expression of early differentiation markers such as keratin 1 in these cells whereas transcription of the late marker loricrin was not affected (Zhu et al., 1999b). The crucial relevance of C/EBP in skin biology and its particular role in keratinocyte maturation was also demonstrated in mice carrying a double knock-out in C/EBPα and C/EBPβ. These animals died a few days after birth as a consequence of tremendous skin dysfunctions such as severe transepithelial water loss and the absence of the lipophilic skin barrier accompanied by a greatly reduced expression of terminal differentiation markers in the epidermis (Lopez et al., 2009).

All together these findings provide strong evidence for the major importance of a precise spatiotemporal control of the MAPK cascades transducing signals from the plasma membrane into the nucleus during proliferation and terminal differentiation of keratinocytes in the skin.

Dysregulated Differentiation in Psoriasis

Dysregulation of the complex differentiation program in keratinocytes can contribute to several pathophysiologies including skin cancer and various dermatoses such as psoriasis (Ambler et al., 2009; Proksch et al., 2008). Psoriasis is a chronic inflammatory disease that is caused by environmental stress and promoted by specific genetic factors (Margadant et al., 2010; Nestle et al., 2009). It is characterized by an inflammatory reaction of the skin and an excessive growth and disturbed maturation of keratinocytes in the epidermis. Studies in psoriatic keratinocytes revealed that the expression of TRPC channels is strongly downregulated in these cells. This results in a diminished Ca^{2+} influx in these cells accompanied by a reduced expression of proteins associated with keratinocyte differentiation such as keratin 10 and transglutaminase 1 (Leuner et al., 2011).

Integrins are in suspect to be key players in the pathophysiology of psoriasis by upregulating the proliferation in keratinocytes. As demonstrated in a mouse model, transgenic expression of β1 integrin caused symptoms of psoriasis in these animals (Carroll et al., 1995). Interestingly, after integrin-mediated activation of ERK1/2 signaling, psoriatic keratinocytes experience hyperproliferation which in turn enhances integrin expression, indicating the existence of a positive feedback loop between integrin production and ERK1/2 stimulation in these cells (Haase et al., 2001; Scholl et al., 2004). The ERK1/2-mediated hyperproliferation was accompanied by a delayed onset of the expression of keratin 10 and involucrin in the subrabasal layers relevant for keratinocyte differentiation in the epidermis (Hobbs et al., 2004). Moreover, psoriasis was reported to be associated with increased levels of p38 and ERK1/2 activity suggesting that both kinases might play a pivotal role in the pathophysiology of this disease (Johansen et al., 2005). Further detailed studies are required for a better understanding of the pathomechanisms contributing to psoriasis.

Unfortunately, recent clinical trials testing the systemic administration of synthetic MAPK inhibitors in the treatment of inflammatory diseases failed due to poor bioavailability and tremendous side effects (Hammaker et al., 2010; Roberts et al., 2007; Wang et al., 2007; Yong et al., 2009). However, psoriasis and other skin diseases provide the opportunity that medication could be topically applied directly to the affected areas of the skin. This might avoid the drawbacks of systemic drug application and thus raises hopes for the development of innovative therapeutical strategies based on the use of synthetic MAPK inhibitors (Hoesel et al., 2009; Medicherla et al., 2009; Medicherla et al., 2010; Pastore et al., 2005).

Conclusion

The continuous renewal of the epidermis is majorly determined by a well balanced regulation of division, migration and maturation of epidermal stem cells present in the stratum basale. The MAPK signaling pathways are key players in the control of proliferation and differentiation in immature keratinocytes. The unraveling of the underlying molecular mechansims however is aggravated by the multiple bidirectional crosstalks between the different MAPK cascades and the complexity of downstream effects and processes influenced by MAPK activities in these cells at different stages of maturation. In addition to MAPKs, further signal transduction pathways **including that of Notch, NFκB and Wnt are also importantly involved in the** growth and reproduction of skin cells, and have been reviewed elsewhere (Ambler et al., 2009; Okuyama et al., 2008; Panelos et al., 2009; Takao et al., 2003).

There is a broad and detailed knowledge about signaling mechanisms controlling the biology and fate of stem cells in the hair follicle of the skin (Fuchs, 2007; Oshima et al., 2001; Taylor et al., 2000). Less work however has been done on the interfollicular stem cells to understand their complex life cycle that essentially contributes to the renewal and regeneration of the epidermis. Thus further investigations are thoroughly required to enhance insight into the molecular mechanisms regulating the growth of epidermal stem cells and their differentiation into terminally differentiated keratinocytes. An improved comprehension of the keratinocyte biology would greatly stimulate the development of target-directed therapeutic strategies for the treatment of epidermal skin pathologies such as psoriasis.

References

Adhikary, G., Chew, Y.C., Reece, E.A., and Eckert, R.L. 2010. PKC-delta and -eta, MEKK-1, MEK-6, MEK-3, and p38-delta are essential mediators of the response of normal human epidermal keratinocytes to differentiating agents. *J. Invest Dermatol.* 130:2017-2030.

Al-Shanti, N. and Stewart, C.E. 2008. PD98059 enhances C2 myoblast differentiation through p38 MAPK activation: a novel role for PD98059. *J. Endocrinol.* 198:243-252.

Alam, H., Sehgal, L., Kundu, S.T., Dalal, S.N., and Vaidya, M.M. 2011. Novel function of keratins 5 and 14 in proliferation and differentiation of stratified epithelial cells. *Mol. Biol. Cell* 22:4068-4078.

Allen, M., Svensson, L., Roach, M., Hambor, J., McNeish, J., and Gabel, C.A. 2000. Deficiency of the stress kinase p38alpha results in embryonic lethality: characterization of the kinase dependence of stress responses of enzyme-deficient embryonic stem cells. *J. Exp. Med.* 191:859-870.

Ambler, C.A. and Maatta, A. 2009. Epidermal stem cells: location, potential and contribution to cancer. *J. Pathol.* 217:206-216.

Bianco, S.D., Peng, J.B., Takanaga, H., Suzuki, Y., Crescenzi, A., Kos, C.H., Zhuang, L., Freeman, M.R., Gouveia, C.H., Wu, J., Luo, H., Mauro, T., Brown, E.M., and Hediger, M.A. 2007. Marked disturbance of calcium homeostasis in mice with targeted disruption of the Trpv6 calcium channel gene. *J. Bone Miner. Res.* 22:274-285.

Blanpain, C. and Fuchs, E. 2009. Epidermal homeostasis: a balancing act of stem cells in the skin. *Nat. Rev. Mol. Cell Biol.* 10:207-217.

Bouschet, T. and Henley, J.M. 2005. Calcium as an extracellular signalling molecule: perspectives on the Calcium Sensing Receptor in the brain. *C. R. Biol.* 328:691-700.

Breitkreutz, D., Braiman-Wiksman, L., Daum, N., Denning, M.F., and Tennenbaum, T. 2007. Protein kinase C family: on the crossroads of cell signaling in skin and tumor epithelium. *J. Cancer Res. Clin. Oncol.* 133:793-808.

Broome, A.M., Ryan, D., and Eckert, R.L. 2003. S100 protein subcellular localization during epidermal differentiation and psoriasis. *J. Histochem. Cytochem.* 51:675-685.

Candi, E., Schmidt, R., and Melino, G. 2005. The cornified envelope: a model of cell death in the skin. *Nat. Rev. Mol. Cell Biol.* 6:328-340.

Carroll, J.M., Romero, M.R., and Watt, F.M. 1995. Suprabasal integrin expression in the epidermis of transgenic mice results in developmental defects and a phenotype resembling psoriasis. *Cell* 83:957-968.

Clayton, E., Doupe, D.P., Klein, A.M., Winton, D.J., Simons, B.D., and Jones, P.H. 2007. A single type of progenitor cell maintains normal epidermis. *Nature* 446:185-189.

Coulombe, P.A. and Lee, C.H. 2012. Defining keratin protein function in skin epithelia: epidermolysis bullosa simplex and its aftermath. *J. Invest Dermatol.* 132:763-775.

Dajee, M., Tarutani, M., Deng, H., Cai, T., and Khavari, P.A. 2002. Epidermal Ras blockade demonstrates spatially localized Ras promotion of proliferation and inhibition of differentiation. Oncogene 21:1527-1538.

Dashti, S.R., Efimova, T., and Eckert, R.L. 2001a. MEK6 regulates human involucrin gene expression via a p38alpha - and p38delta -dependent mechanism. *J. Biol. Chem.* 276:27214-27220.

Dashti, S.R., Efimova, T., and Eckert, R.L. 2001b. MEK7-dependent activation of p38 MAP kinase in keratinocytes. *J. Biol. Chem.* 276:8059-8063.

Denning, M.F. 2004. Epidermal keratinocytes: regulation of multiple cell phenotypes by multiple protein kinase C isoforms. *Int. J. Biochem. Cell Biol.* 36:1141-1146.

Denning, M.F. 2010. Protein kinase C/mitogen-activated protein kinase signaling in keratinocyte differentiation control. *J. Invest Dermatol.* 130:1968-1970.

Denning, M.F., Dlugosz, A.A., Williams, E.K., Szallasi, Z., Blumberg, P.M., and Yuspa, S.H. 1995. Specific protein kinase C isozymes mediate the induction of keratinocyte differentiation markers by calcium. *Cell Growth Differ.* 6:149-157.

Donato, R. 2001. S100: a multigenic family of calcium-modulated proteins of the EF-hand type with intracellular and extracellular functional roles. *Int. J. Biochem. Cell Biol.* 33:637-668.

Doupe, D.P. and Jones, P.H. 2012. Interfollicular epidermal homeostasis: dicing with differentiation. *Exp. Dermatol.* 21:249-253.

Eckert, R.L., Broome, A.M., Ruse, M., Robinson, N., Ryan, D., and Lee, K. 2004a. S100 proteins in the epidermis. *J. Invest Dermatol.* 123:23-33.

Eckert, R.L., Crish, J.F., Efimova, T., Dashti, S.R., Deucher, A., Bone, F., Adhikary, G., Huang, G., Gopalakrishnan, R., and Balasubramanian, S. 2004b. Regulation of involucrin gene expression. *J. Invest Dermatol.* 123:13-22.

Eckert, R.L. and Green, H. 1986. Structure and evolution of the human involucrin gene. *Cell* 46:583-589.

Eckert, R.L., Sturniolo, M.T., Broome, A.M., Ruse, M., and Rorke, E.A. 2005. Transglutaminase function in epidermis. *J. Invest Dermatol.* 124:481-492.

Efimova, T., Broome, A.M., and Eckert, R.L. 2003. A regulatory role for p38 delta MAPK in keratinocyte differentiation. Evidence for p38 delta-ERK1/2 complex formation. *J. Biol. Chem.* 278:34277-34285.

Efimova, T., Deucher, A., Kuroki, T., Ohba, M., and Eckert, R.L. 2002. Novel protein kinase C isoforms regulate human keratinocyte differentiation by activating a p38 delta mitogen-activated protein kinase cascade that targets CCAAT/enhancer-binding protein alpha. *J. Biol. Chem.* 277:31753-31760.

Elias, P.M., Ahn, S.K., Denda, M., Brown, B.E., Crumrine, D., Kimutai, L.K., Komuves, L., Lee, S.H., and Feingold, K.R. 2002. Modulations in epidermal calcium regulate the expression of differentiation-specific markers. *J. Invest Dermatol.* 119:1128-1136.

Epstein, W.L. and Maibach, H.I. 1965. Cell renewal in human epidermis. *Arch. Dermatol.* 92:462-468.

Fey, D., Croucher, D.R., Kolch, W., and Kholodenko, B.N. 2012. Crosstalk and signaling switches in mitogen-activated protein kinase cascades. *Front Physiol* 3:355-

Fisher, G.J., Tavakkol, A., Leach, K., Burns, D., Basta, P., Loomis, C., Griffiths, C.E., Cooper, K.D., Reynolds, N.J., Elder, J.T. 1993. Differential expression of protein kinase C isoenzymes in normal and psoriatic adult human skin: reduced expression of protein kinase C-beta II in psoriasis. *J. Invest Dermatol.* 101:553-559.

Fuchs, E. 2007. Scratching the surface of skin development. *Nature* 445:834-842.

Fuchs, E. and Chen, T. 2012. A matter of life and death: self-renewal in stem cells. *EMBO Rep.*

Fuchs, E. and Green, H. 1980. Changes in keratin gene expression during terminal differentiation of the keratinocyte. *Cell* 19:1033-1042.

Gazel, A., Banno, T., Walsh, R., and Blumenberg, M. 2006. Inhibition of JNK promotes differentiation of epidermal keratinocytes. *J. Biol. Chem.* 281:20530-20541.

Gazel, A., Nijhawan, R.I., Walsh, R., and Blumenberg, M. 2008. Transcriptional profiling defines the roles of ERK and p38 kinases in epidermal keratinocytes. *J. Cell Physiol* 215:292-308.

Goldsmith, L.A. 1990. My organ is bigger than your organ. *Arch. Dermatol.* 126:301-302.

Haase, I., Hobbs, R.M., Romero, M.R., Broad, S., and Watt, F.M. 2001. A role for mitogen-activated protein kinase activation by integrins in the pathogenesis of psoriasis. *J. Clin. Invest* 108:527-536.

Halprin, K.M. 1972. Epidermal "turnover time"--a re-examination. *Br. J. Dermatol.* 86:14-19.

Hammaker, D. and Firestein, G.S. 2010. "Go upstream, young man": lessons learned from the p38 saga. *Ann. Rheum. Dis.* 69 Suppl 1:i77-i82.

Han, B., Rorke, E.A., Adhikary, G., Chew, Y.C., Xu, W., and Eckert, R.L. 2012. Suppression of AP1 transcription factor function in keratinocyte suppresses differentiation. *PLoS. One.* 7:e36941-

Hara, T., Miyazaki, M., Hakuno, F., Takahashi, S., and Chida, K. 2011. PKCeta promotes a proliferation to differentiation switch in keratinocytes via upregulation of p27Kip1 mRNA through suppression of JNK/c-Jun signaling under stress conditions. *Cell Death. Dis.* 2:e157-

Heinrichsdorff, J., Luedde, T., Perdiguero, E., Nebreda, A.R., and Pasparakis, M. 2008. p38 alpha MAPK inhibits JNK activation and collaborates with IkappaB kinase 2 to prevent endotoxin-induced liver failure. *EMBO Rep.* 9:1048-1054.

Hitomi, K. 2005. Transglutaminases in skin epidermis. *Eur. J. Dermatol.* 15:313-319.

Hobbs, R.M., Silva-Vargas, V., Groves, R., and Watt, F.M. 2004. Expression of activated MEK1 in differentiating epidermal cells is sufficient to generate hyperproliferative and inflammatory skin lesions. *J. Invest Dermatol.* 123:503-515.

Hoesel, L.M., Mattar, A.F., Arbabi, S., Niederbichler, A.D., Ipaktchi, K., Su, G.L., Westfall, M.V., Wang, S.C., and Hemmila, M.R. 2009. Local wound p38 MAPK inhibition attenuates burn-induced cardiac dysfunction. *Surgery* 146:775-785.

Hohl, D., Mehrel, T., Lichti, U., Turner, M.L., Roop, D.R., and Steinert, P.M. 1991. Characterization of human loricrin. Structure and function of a new class of epidermal cell envelope proteins. *J. Biol. Chem.* 266:6626-6636.

Hui, L., Bakiri, L., Mairhorfer, A., Schweifer, N., Haslinger, C., Kenner, L., Komnenovic, V., Scheuch, H., Beug, H., and Wagner, E.F. 2007. p38alpha suppresses normal and cancer cell proliferation by antagonizing the JNK-c-Jun pathway. *Nat. Genet.* 39:741-749.

Iizuka, R., Chiba, K., and Imajoh-Ohmi, S. 2003. A novel approach for the detection of proteolytically activated transglutaminase 1 in epidermis using cleavage site-directed antibodies. *J. Invest Dermatol.* 121:457-464.

Jans, R., Atanasova, G., Jadot, M., and Poumay, Y. 2004. Cholesterol depletion upregulates involucrin expression in epidermal keratinocytes through activation of p38. *J. Invest Dermatol.* 123:564-573.

Jensen, J.M., Folster-Holst, R., Baranowsky, A., Schunck, M., Winoto-Morbach, S., Neumann, C., Schutze, S., and Proksch, E. 2004. Impaired sphingomyelinase activity and epidermal differentiation in atopic dermatitis. *J. Invest Dermatol.* 122:1423-1431.

Johansen, C., Kragballe, K., Westergaard, M., Henningsen, J., Kristiansen, K., and Iversen, L. 2005. The mitogen-activated protein kinases p38 and ERK1/2 are increased in lesional psoriatic skin. *Br. J. Dermatol.* 152:37-42.

Jonak, C., Mildner, M., Klosner, G., Paulitschke, V., Kunstfeld, R., Pehamberger, H., Tschachler, E., and Trautinger, F. 2011. The hsp27kD heat shock protein and p38-MAPK signaling are required for regular epidermal differentiation. *J. Dermatol. Sci.* 61:32-37.

Jones, P.H. and Watt, F.M. 1993. Separation of human epidermal stem cells from transit amplifying cells on the basis of differences in integrin function and expression. *Cell* 73:713-724.

Jungersted, J.M., Hellgren, L.I., Jemec, G.B., and Agner, T. 2008. Lipids and skin barrier function--a clinical perspective. *Contact Dermatitis* 58:255-262.

Kawachi, Y., Ishitsuka, Y., Maruyama, H., Fujisawa, Y., Furuta, J., Nakamura, Y., and Otsuka, F. 2012. GATA-3 regulates differentiation-specific loricrin gene expression in keratinocytes. *Exp. Dermatol.* 21:859-864.

Kayahara, M., Wang, X., and Tournier, C. 2005. Selective regulation of c-jun gene expression by mitogen-activated protein kinases via the 12-o-tetradecanoylphorbol-13-acetate- responsive element and myocyte enhancer factor 2 binding sites. *Mol. Cell Biol.* 25:3784-3792.

Kuwano, Y., Fujimoto, M., Watanabe, R., Ishiura, N., Nakashima, H., Komine, M., Hamazaki, T.S., Tamaki, K., and Okochi, H. 2007. The involvement of Gab1 and PI 3-kinase in beta1 integrin signaling in keratinocytes. *Biochem. Biophys. Res. Commun.* 361:224-229.

Kypriotou, M., Huber, M., and Hohl, D. 2012. The human epidermal differentiation complex: cornified envelope precursors, S100 proteins and the 'fused genes' family. *Exp. Dermatol.* 21:643-649.

Lane, E.B. and McLean, W.H. 2004. Keratins and skin disorders. *J. Pathol.* 204:355-366.

Lechler, T. and Fuchs, E. 2005. Asymmetric cell divisions promote stratification and differentiation of mammalian skin. *Nature* 437:275-280.

Lee, J.W. and Juliano, R. 2004. Mitogenic signal transduction by integrin- and growth factor receptor-mediated pathways. *Mol. Cells* 17:188-202.

Legewie, S., Bluthgen, N., Schafer, R., and Herzel, H. 2005. Ultrasensitization: switch-like regulation of cellular signaling by transcriptional induction. *PLoS. Comput. Biol.* 1:e54-

Lehen'kyi, V., Beck, B., Polakowska, R., Charveron, M., Bordat, P., Skryma, R., and Prevarskaya, N. 2007. TRPV6 is a Ca2+ entry channel essential for Ca2+-induced differentiation of human keratinocytes. *J. Biol. Chem.* 282:22582-22591.

Leuner, K., Kraus, M., Woelfle, U., Beschmann, H., Harteneck, C., Boehncke, W.H., Schempp, C.M., and Muller, W.E. 2011. Reduced TRPC channel expression in psoriatic keratinocytes is associated with impaired differentiation and enhanced proliferation. *PLoS. One.* 6:e14716-

Levy, L., Broad, S., Diekmann, D., Evans, R.D., and Watt, F.M. 2000. beta1 integrins regulate keratinocyte adhesion and differentiation by distinct mechanisms. *Mol. Biol. Cell* 11:453-466.

Levy, V., Lindon, C., Zheng, Y., Harfe, B.D., and Morgan, B.A. 2007. Epidermal stem cells arise from the hair follicle after wounding. *FASEB J.* 21:1358-1366.

Litjens, S.H., de Pereda, J.M., and Sonnenberg, A. 2006. Current insights into the formation and breakdown of hemidesmosomes. *Trends Cell Biol.* 16:376-383.

Lopez, R.G., Garcia-Silva, S., Moore, S.J., Bereshchenko, O., Martinez-Cruz, A.B., Ermakova, O., Kurz, E., Paramio, J.M., and Nerlov, C. 2009. C/EBPalpha and beta couple interfollicular keratinocyte proliferation arrest to commitment and terminal differentiation. *Nat. Cell Biol.* 11:1181-1190.

Marekov, L.N. and Steinert, P.M. 1998. Ceramides are bound to structural proteins of the human foreskin epidermal cornified cell envelope. *J. Biol. Chem.* 273:17763-17770.

Margadant, C., Charafeddine, R.A., and Sonnenberg, A. 2010. Unique and redundant functions of integrins in the epidermis. *FASEB J.* 24:4133-4152.

Marinkovich, M.P., Keene, D.R., Rimberg, C.S., and Burgeson, R.E. 1993. Cellular origin of the dermal-epidermal basement membrane. *Dev. Dyn.* 197:255-267.

Marques-Pereira, J.P. and Leblond, C.P. 2012. Mitosis and differentiation in the stratified squamous epithelium of the rat esophagus. *Am. J. Anat.* 117:73-87.

Martin, P. 1997. Wound healing--aiming for perfect skin regeneration. *Science* 276:75-81.

Mascre, G., Dekoninck, S., Drogat, B., Youssef, K.K., Brohee, S., Sotiropoulou, P.A., Simons, B.D., and Blanpain, C. 2012. Distinct contribution of stem and progenitor cells to epidermal maintenance. *Nature* 489:257-262.

Matsuki, M., Yamashita, F., Ishida-Yamamoto, A., Yamada, K., Kinoshita, C., Fushiki, S., Ueda, E., Morishima, Y., Tabata, K., Yasuno, H., Hashida, M., Iizuka, H., Ikawa, M., Okabe, M., Kondoh, G., Kinoshita, T., Takeda, J., and Yamanishi, K. 1998. Defective stratum corneum and early neonatal death in mice lacking the gene for transglutaminase 1 (keratinocyte transglutaminase). *Proc. Natl. Acad. Sci. U. S. A* 95:1044-1049.

Matsumoto, M., Sudo, T., Saito, T., Osada, H., and Tsujimoto, M. 2000. Involvement of p38 mitogen-activated protein kinase signaling pathway in osteoclastogenesis mediated by receptor activator of NF-kappa B ligand (RANKL). *J. Biol. Chem.* 275:31155-31161.

McKay, M.M. and Morrison, D.K. 2007. Integrating signals from RTKs to ERK/MAPK. *Oncogene* 26:3113-3121.

Medicherla, S., Ma, J.Y., Reddy, M., Esikova, I., Kerr, I., Movius, F., Higgins, L.S., and Protter, A.A. 2010. Topical alpha-selective p38 MAP kinase inhibition reduces acute skin inflammation in guinea pig. *J. Inflamm. Res.* 3:9-16.

Medicherla, S., Wadsworth, S., Cullen, B., Silcock, D., Ma, J.Y., Mangadu, R., Kerr, I., Chakravarty, S., Luedtke, G.L., Dugar, S., Protter, A.A., and Higgins, L.S. 2009. p38 MAPK inhibition reduces diabetes-induced impairment of wound healing. *Diabetes Metab Syndr. Obes.* 2:91-100.

Michel, S., Bernerd, F., Jetten, A.M., Floyd, E.E., Shroot, B., and Reichert, U. 1992. Expression of keratinocyte transglutamine mRNA revealed by in situ hybridization. *J. Invest Dermatol.* 98:364-368.

Muller, M., Essin, K., Hill, K., Beschmann, H., Rubant, S., Schempp, C.M., Gollasch, M., Boehncke, W.H., Harteneck, C., Muller, W.E., and Leuner, K. 2008. Specific TRPC6 channel activation, a novel approach to stimulate keratinocyte differentiation. *J. Biol. Chem.* 283:33942-33954.

Myllyharju, J. and Kivirikko, K.I. 2004. Collagens, modifying enzymes and their mutations in humans, flies and worms. *Trends Genet.* 20:33-43.

Nakagawa, H. and Maeda, S. 2012. Molecular mechanisms of liver injury and hepatocarcinogenesis: focusing on the role of stress-activated MAPK. *Patholog. Res. Int.* 2012:172894-

Nestle, F.O., Kaplan, D.H., and Barker, J. 2009. Psoriasis. *N. Engl. J. Med.* 361:496-509.

Okuyama, R., Tagami, H., and Aiba, S. 2008. Notch signaling: its role in epidermal homeostasis and in the pathogenesis of skin diseases. *J. Dermatol. Sci.* 49:187-194.

Oshima, H., Rochat, A., Kedzia, C., Kobayashi, K., and Barrandon, Y. 2001. Morphogenesis and renewal of hair follicles from adult multipotent stem cells. *Cell* 104:233-245.

Panelos, J. and Massi, D. 2009. Emerging role of Notch signaling in epidermal differentiation and skin cancer. *Cancer Biol. Ther.* 8:1986-1993.

Park, C.H. and Chung, J.H. 2011. Epidermal growth factor-induced matrix metalloproteinase-1 expression is negatively regulated by p38 MAPK in human skin fibroblasts. *J. Dermatol. Sci.* 64:134-141.

Pastore, S., Mascia, F., Mariotti, F., Dattilo, C., Mariani, V., and Girolomoni, G. 2005. ERK1/2 regulates epidermal chemokine expression and skin inflammation. *J. Immunol.* 174:5047-5056.

Pearson, G., Robinson, F., Beers, G.T., Xu, B.E., Karandikar, M., Berman, K., and Cobb, M.H. 2001. Mitogen-activated protein (MAP) kinase pathways: regulation and physiological functions. *Endocr. Rev.* 22:153-183.

Peng, J., Li, W., Li, H., Jia, Y., and Liu, Z. 2009. Inhibition of p38 MAPK facilitates ex vivo expansion of skin epithelial progenitor cells. *In Vitro Cell Dev. Biol. Anim* 45:558-565.

Popp, T., Egea, V., Kehe, K., Steinritz, D., Schmidt, A., Jochum, M., and Ries, C. 2011. Sulfur mustard induces differentiation in human primary keratinocytes: opposite roles of p38 and ERK1/2 MAPK. *Toxicol. Lett.* 204:43-51.

Porter, R.M. and Lane, E.B. 2003. Phenotypes, genotypes and their contribution to understanding keratin function. *Trends Genet.* 19:278-285.

Potten, C.S. 1974. The epidermal proliferative unit: the possible role of the central basal cell. *Cell Tissue Kinet.* 7:77-88.

Proksch, E., Brandner, J.M., and Jensen, J.M. 2008. The skin: an indispensable barrier. *Exp. Dermatol.* 17:1063-1072.

Ramsey, I.S., Delling, M., and Clapham, D.E. 2006. An introduction to TRP channels. *Annu. Rev. Physiol* 68:619-647.

Rawlings, A.V., Scott, I.R., Harding, C.R., and Bowser, P.A. 1994. Stratum corneum moisturization at the molecular level. J. Invest Dermatol. 103:731-741.

Roberts, P.J. and Der, C.J. 2007. Targeting the Raf-MEK-ERK mitogen-activated protein kinase cascade for the treatment of cancer. *Oncogene* 26:3291-3310.

Robinson, N.A., Lapic, S., Welter, J.F., and Eckert, R.L. 1997. S100A11, S100A10, annexin I, desmosomal proteins, small proline-rich proteins, plasminogen activator inhibitor-2, and involucrin are components of the cornified envelope of cultured human epidermal keratinocytes. *J. Biol. Chem.* 272:12035-12046.

Roper, E., Weinberg, W., Watt, F.M., and Land, H. 2001. p19ARF-independent induction of p53 and cell cycle arrest by Raf in murine keratinocytes. *EMBO Rep.* 2:145-150.

Rorke, E.A., Adhikary, G., Jans, R., Crish, J.F., and Eckert, R.L. 2010. AP1 factor inactivation in the suprabasal epidermis causes increased epidermal hyperproliferation and hyperkeratosis but reduced carcinogen-dependent tumor formation. *Oncogene* 29:5873-5882.

Rossi, A., Jang, S.I., Ceci, R., Steinert, P.M., and Markova, N.G. 1998. Effect of AP1 transcription factors on the regulation of transcription in normal human epidermal keratinocytes. *J. Invest Dermatol.* 110:34-40.

Ruse, M., Lambert, A., Robinson, N., Ryan, D., Shon, K.J., and Eckert, R.L. 2001. S100A7, S100A10, and S100A11 are transglutaminase substrates. *Biochemistry* 40:3167-3173.

Schmidt, M., Goebeler, M., Posern, G., Feller, S.M., Seitz, C.S., Brocker, E.B., Rapp, U.R., and Ludwig, S. 2000. Ras-independent activation of the Raf/MEK/ERK pathway upon calcium-induced differentiation of keratinocytes. *J. Biol. Chem.* 275:41011-41017.

Scholl, F.A., Dumesic, P.A., and Khavari, P.A. 2004. Mek1 alters epidermal growth and differentiation. *Cancer Res.* 64:6035-6040.

Schultz, G.S. and Wysocki, A. 2009. Interactions between extracellular matrix and growth factors in wound healing. *Wound. Repair Regen.* 17:153-162.

Seo, H.R., Kwan, Y.W., Cho, C.K., Bae, S., Lee, S.J., Soh, J.W., Chung, H.Y., and Lee, Y.S. 2004. PKCalpha induces differentiation through ERK1/2 phosphorylation in mouse keratinocytes. *Exp. Mol. Med.* 36:292-299.

Shibata, S., Tada, Y., Asano, Y., Hau, C.S., Kato, T., Saeki, H., Yamauchi, T., Kubota, N., Kadowaki, T., and Sato, S. 2012. Adiponectin regulates cutaneous wound healing by promoting keratinocyte proliferation and migration via the ERK signaling pathway. *J. Immunol.* 189:3231-3241.

Silvers, A.L., Bachelor, M.A., and Bowden, G.T. 2003. The role of JNK and p38 MAPK activities in UVA-induced signaling pathways leading to AP-1 activation and c-Fos expression. *Neoplasia.* 5:319-329.

Simons, K. and Toomre, D. 2000. Lipid rafts and signal transduction. *Nat. Rev. Mol. Cell Biol.* 1:31-39.

Simpson, C.L., Patel, D.M., and Green, K.J. 2011. Deconstructing the skin: cytoarchitectural determinants of epidermal morphogenesis. *Nat. Rev. Mol. Cell Biol.* 12:565-580.

Smith, E.L., Walworth, N.C., and Holick, M.F. 1986. Effect of 1 alpha,25-dihydroxyvitamin D3 on the morphologic and biochemical differentiation of cultured human epidermal keratinocytes grown in serum-free conditions. *J. Invest Dermatol.* 86:709-714.

Steinert, P.M., Cantieri, J.S., Teller, D.C., Lonsdale-Eccles, J.D., and Dale, B.A. 1981. Characterization of a class of cationic proteins that specifically interact with intermediate filaments. *Proc. Natl. Acad. Sci. U. S. A* 78:4097-4101.

Steinert, P.M., Kartasova, T., and Marekov, L.N. 1998. Biochemical evidence that small proline-rich proteins and trichohyalin function in epithelia by modulation of the biomechanical properties of their cornified cell envelopes. *J. Biol. Chem.* 273:11758-11769.

Su, M.J., Bikle, D.D., Mancianti, M.L., and Pillai, S. 1994. 1,25-Dihydroxyvitamin D3 potentiates the keratinocyte response to calcium. *J. Biol. Chem.* 269:14723-14729.

Takao, J., Yudate, T., Das, A., Shikano, S., Bonkobara, M., Ariizumi, K., and Cruz, P.D. 2003. Expression of NF-kappaB in epidermis and the relationship between NF-kappaB activation and inhibition of keratinocyte growth. *Br. J. Dermatol.* 148:680-688.

Taylor, G., Lehrer, M.S., Jensen, P.J., Sun, T.T., and Lavker, R.M. 2000. Involvement of follicular stem cells in forming not only the follicle but also the epidermis. *Cell* 102:451-461.

Tesfaigzi, J. and Carlson, D.M. 1999. Expression, regulation, and function of the SPR family of proteins. A review. *Cell Biochem. Biophys.* 30:243-265.

Thalhamer, T., McGrath, M.A., and Harnett, M.M. 2008. MAPKs and their relevance to arthritis and inflammation. *Rheumatology.* (Oxford) 47:409-414.

Tsuji, T., Kitajima, S., and Koashi, Y. 1998. Expression of proliferating cell nuclear antigen (PCNA) and apoptosis related antigen (LeY) in epithelial skin tumors. *Am. J. Dermatopathol.* 20:164-169.

Tu, C.L., Crumrine, D.A., Man, M.Q., Chang, W., Elalieh, H., You, M., Elias, P.M., and Bikle, D.D. 2012. Ablation of the calcium-sensing receptor in keratinocytes impairs epidermal differentiation and barrier function. *J. Invest Dermatol.* 132:2350-2359.

Usui, M.L., Mansbridge, J.N., Carter, W.G., Fujita, M., and Olerud, J.E. 2008. Keratinocyte migration, proliferation, and differentiation in chronic ulcers from patients with diabetes and normal wounds. *J. Histochem. Cytochem.* 56:687-696.

Volz, A., Korge, B.P., Compton, J.G., Ziegler, A., Steinert, P.M., and Mischke, D. 1993. Physical mapping of a functional cluster of epidermal differentiation genes on chromosome 1q21. *Genomics* 18:92-99.

Vu, T.H. and Werb, Z. 2000. Matrix metalloproteinases: effectors of development and normal physiology. *Genes Dev.* 14:2123-2133.

Wang, D., Boerner, S.A., Winkler, J.D., and LoRusso, P.M. 2007. Clinical experience of MEK inhibitors in cancer therapy. *Biochim. Biophys. Acta* 1773:1248-1255.

Wang, Z., Yang, H., Tachado, S.D., Capo-Aponte, J.E., Bildin, V.N., Koziel, H., and Reinach, P.S. 2006. Phosphatase-mediated crosstalk control of ERK and p38 MAPK signaling in corneal epithelial cells. *Invest Ophthalmol. Vis. Sci.* 47:5267-5275.

Watt, F.M. and Jensen, K.B. 2009. Epidermal stem cell diversity and quiescence. *EMBO Mol. Med.* 1:260-267.

Wolfman, J.C., Palmby, T., Der, C.J., and Wolfman, A. 2002. Cellular N-Ras promotes cell survival by downregulation of Jun N-terminal protein kinase and p38. *Mol. Cell Biol.* 22:1589-1606.

Yamane, K., Jinnin, M., Etoh, T., Kobayashi, Y., Shimozono, N., Fukushima, S., Masuguchi, S., Maruo, K., Inoue, Y., Ishihara, T., Aoi, J., Oike, Y., and Ihn, H. 2012. Down-regulation of miR-124/-214 in cutaneous squamous cell carcinoma mediates abnormal cell proliferation via the induction of ERK. *J. Mol. Med.* (Berl)

Yang, L.C., Ng, D.C., and Bikle, D.D. 2003. Role of protein kinase C alpha in calcium induced keratinocyte differentiation: defective regulation in squamous cell carcinoma. *J. Cell Physiol* 195:249-259.

Yong, H.Y., Koh, M.S., and Moon, A. 2009. The p38 MAPK inhibitors for the treatment of inflammatory diseases and cancer. *Expert. Opin. Investig. Drugs* 18:1893-1905.

Yuan, Z.Q., Feldman, R.I., Sussman, G.E., Coppola, D., Nicosia, S.V., and Cheng, J.Q. 2003. AKT2 inhibition of cisplatin-induced JNK/p38 and Bax activation by phosphorylation of ASK1: implication of AKT2 in chemoresistance. *J. Biol. Chem.* 278:23432-23440.

Zhang, H., Shi, X., Hampong, M., Blanis, L., and Pelech, S. 2001. Stress-induced inhibition of ERK1 and ERK2 by direct interaction with p38 MAP kinase. *J. Biol. Chem.* 276:6905-6908.

Zhu, A.J., Haase, I., and Watt, F.M. 1999a. Signaling via beta1 integrins and mitogen-activated protein kinase determines human epidermal stem cell fate in vitro. *Proc. Natl. Acad. Sci. U. S. A* 96:6728-6733.

Zhu, S., Oh, H.S., Shim, M., Sterneck, E., Johnson, P.F., and Smart, R.C. 1999b. C/EBPbeta modulates the early events of keratinocyte differentiation involving growth arrest and keratin 1 and keratin 10 expression. *Mol. Cell Biol.* 19:7181-7190.

Zimmer, D.B., Wright, S.P., and Weber, D.J. 2003. Molecular mechanisms of S100-target protein interactions. *Microsc. Res. Tech.* 60:552-559.

In: Keratinocytes
Editor: Elia Ranzato

ISBN: 978-1-62618-798-6
© 2013 Nova Science Publishers, Inc.

Chapter 4

Epithelial-to-Mesenchymal Transition in Epidermal Keratinocytes

Ekaterina A. Vorotelyak[1],*, *Vasily V. Terskikh*[1]
and Andrey V. Vasiliev[1]
[1]N. K. Koltsov Institute of Developmental Biology,
Russian Academy of Sciences, Moscow, Russia

Abstract

Epithelium and mesenchyme are the two main phenotypes that contrast each other in the organization of the cytoskeleton, metabolic apparatus and cell behavior. Epithelium-to-mesenchyme transition (EMT) and the reverse process, termed mesenchyme-to-epithelium transition (MET), play central developmental roles in the development of the body plan's formation and in the cell differentiation of multiple tissues and organs. EMT is a biological process that occurs when polarized epithelial cells assume a mesenchymal cell phenotype, characterized by an enhanced migratory capacity; a cytoskeleton reorganization to confer the ability to move through a three-dimensional ECM; an elevated resistance to apoptosis; morphogenetic events; an increased production of ECM components; and a new transcriptional program, which is induced to

---

* E-mail: vorotelyak@yandex.ru.

maintain the mesenchymal phenotype. Factors and some mechanisms of EMT are considered. The epidermal cell phenotype of an intact epidermis dramatically changes after wounding or inoculation of cells into culture. Keratinocytes undergo activation resulting in elevated proliferation and changes of keratin expression and cell-cell adhesion, gaining mesenchymal features including enhanced motility and cytoskeletal rearrangements.

Cultured keratinocytes exhibit signs of partial EMT. Migrating leading cells with the intermediate EMT phenotype are also typical for wound healing *in vitro* and *in vivo*. Morphogenesis in cultured keratinocytes may be manifested by the formation of tubules under the influence of certain factors including growth factors and culture mediums conditioned by dermal papilla cells. Leading tubular cells invading collagen gel resemble cells situated on the leading edge of migrating multicellular sheets. The authors discuss:epidermal cell plasticity and EMT associated with wound healing and the collective migration of cells in colonies, collagen gel invasion and the formation of tubule-like structures in cultured human epidermal keratinocytes. Cellular and molecular mechanisms of collective migration and tubulogenesis are considered. Tubulogenesis, a morphogenetic process situated at the crossroads of developmental and cell biology, arises from different epithelila and endothelial cells and can be induced by several signaling pathways which converge on distinct downstream events of EMT. Cultured epithelial cells can assume various phenotypic states that recapitulate some specific steps of embryogenesis when cells appear to be plastic and are able to move back and forth between epithelial and mesenchymal states. EMT and the plasticity of epidermal cells are of paramount importance in skin development and regenerative medicine when cultured keratinocyte sheets suitable for grafting on burn wounds are produced.

Introduction

Epithelium and mesenchyme are the two main phenotypes that greatly contrast each other in the organization of the cytoskeleton, metabolic apparatus and cell behavior. Epithelium arises first in embryonic development and mesenchyme arises from the latter through epithelium-to-mesenchyme transition (EMT). EMT and the reverse process, termed mesenchyme-to-epithelium transition (MET), play central roles in development by formating the body plan and in the cell differentiation of multiple tissues and organs (Savagner, 2010).

EMT is a classic example of epithelial cell plasticity. During embryonic development, cells can convert between the epithelial and mesenchymal states through mutual inductions and interactions. Since epithelia typically serve specialized functions, it is assumed that a state of terminal differentiation is protected once development is complete. However, it was proposed that adult epithelia remain responsive to the cues from connective tissue (Cunha, 1985). Now, there is a growing body of evidence showing that epithelial and mesenchymal cell phenotypes are reversible. During embryogenesis, the cells within certain epithelia become increasingly plastic and are able to move back and forth between epithelial and mesenchymal states (Lee et al., 2006). EMT is an important mechanism of mammalian development, and it also contributes to the pathogenesis of diseases such as metastatic cancer and tissue fibrosis.

The Concept of EMT

The concept of EMT was first defined by Elizabeth Hay (Hay, 1968). When studying the formation of the chick primitive streak, which is one of the first signs of gastrulation, Hay proposed that epithelial cells can undergo a **dramatic phenotypic "transformation" (now the term "transition" is used) to** mesenchymal cells. The embryonic program of EMT is a highly coordinated, multistep process that involves profound phenotypic changes such as the down-modulation of cell-cell adhesion structures, including adherens junctions and desmosomes; the loss of cell polarity; the reorganization of the cytoskeleton when intermediate filaments typically switch from cytokeratins to vimentin; the acquisition of migratory and invasive properties, and a resistance to apoptosis and anoikis (Thiery et al. 2009). Epithelial cells can change their phenotype via EMT programs, which enable the cell conversion and production of mesenchymal derivatives during development and adulthood (Neilson, 2007). These mesenchymal–like cells have front end-back end polarity and form only transient contacts with neighbors. They leave the epithelium of origin and move through the underlying matrix. In some situations, colocalizations of distinct markers of epithelial and mesenchymal cells define an intermediate phenotype, indicating cells that have passed only partly through EMT (Jordan et al., 2011). A later EMT concept was extended to the physiological process of partial EMT, which takes place during wound healing and mammary tubulogenesis (Arnoux et al., 2004). EMTs were classified into three different biological subtypes based on the biological context in which they occur (Kalluri & Weinberg, 2009; Zeisberg & Neilson,

2009). Primary, type 1 EMTs occur early during embryonic development and are associated with implantation, embryo formation, and organ development (Sawyer & Fallon, 1983; Sauka-Spengler & Bronner-Fraser, 2008; Thiery & Sleeman, 2006; Acloque et al., 2009). During development, tissues may undergo several rounds of EMT and MET. For example, the heart is formed through three successive cycles of EMT and MET (Nakajima et al., 2000).

In contrast to type 1 EMTs, the type 2 EMTs include physiological processes of partial EMT occurring during wound healing, tissue regeneration, and organ fibrosis. At the wound edges keratinocytes form a hyperproliferative epithelium, which migrates and covers the wounded area with new cells. It is thought that type 1 EMT involving primitive epithelial cells produce mesenchymal cells, whereas type 2 EMT in adult epithelia results in fibroblasts. The best-studied example of the type 2 EMT is associated with renal interstitial fibrosis when, under pathologic conditions, renal tubular epithelial cells can undergo EMT (Okada et al., 1996; Iwano et al., 2002). Myofibroblast activation is a key event playing a critical role in the progression of chronic renal disease (Yang & Liu, 2001). Yang and Liu studied EMT in human tubular epithelial cells incubated with TGF β1. The authors suggested that EMT in renal interstitial fibrosis is a highly regulated process which may be divided into four key steps: 1) loss of epithelial cell adhesion, 2) *de novo* α–smooth musle actin expression and actin reorganization, 3) disruption of tubular basement membrane, and 4) enhanced cell migration and invasion (Yang & Liu, 2001).

Type 3 EMT occurs during tumor progression and malignant transformation, endowing the incipient cancer cells with invasive and metastatic properties (Larue & Bellacosa, 2005; Boyer et al., 1989). While these three classes of EMTs represent distinct biological processes, common genetic and biochemical events appear to support these diverse phenotypic programs (Kalluri & Weinberg, 2009).

EMT is a long-term continuous process, and there is no definitive timepoint at which radical changes take place. Divergence in the gene expression pattern is observed as early as 6 hours after induction of EMT, when more than 60% of all differentially expressed genes are up-regulated, and morphological changes appear gradually over several days (Venkov et al., 2011). The microenvironment signals that induce EMT are associated with *de novo* epigenetic alterations in targeted cells, particulary the hypermethylation of the E-cadherin promoter (Dumont et al., 2008).

In type 3 EMT, major signaling pathways, such as TGF-β, **Wnt**, **Notch**, and Hedgehog, are involved. These pathways converge on several transcription factors, including zinc finger proteins Snai 1 and Snai 2, Twist, ZEB 1/2, and Smads. These factors interact with one another and others to provide crosstalk between the relevant signaling pathways. MicroRNA suppression and epigenetic changes also influence the changes involved in EMT (Talbot et al., 2012). For example, the Snai 1 transcription factor, which is the key regulator of EMT, produces deacetylation of histones H3 and H4 at the E-cadherin promoter and mediates the repression of E-cadherin by recruiting chromatin-modifying activities, forming a multimolecular complex (Peinado et al., 2004). A rapidly increasing body of studies link EMT to epigenetic mechanisms in normal processes, as well as abnormal transitions that lead to oncogenesis. Aberrant changes in epigenetic mechanisms such as DNA methylation, histone modifications and expression of micro RNAs play an important role in cancer and contribute to malignant transitions (McDonald et al., 2011; Stadler & Allis, 2012; Botchkarev et al., 2012). A normal intact epithelium does not typically exhibit features of an EMT. However, under the influence of the surrounding microenvironment, epithelial cells can acquire mesenchymal phenotype. Some data shows that senescent fibroblasts can have deleterious effects on the tissue's **microenvironment and induce EMT** (Laberge et al., 2012).

Molecular Hallmarks of EMT

EMT is characterized by multiple morphological, metabolic, and behavioral changes in epithelial cells that enable them to assume a mesenchymal cell phenotype, which is characterized by enhanced migratory capacity, invasiveness, greatly increased production of ECM components, and resistance to apoptosis (Kalluri & Neilson, 2003). While EMT may be induced by a plethora of signaling pathways, several molecular mechanisms contribute directly to the loss of the epithelial phenotype. They include the activation of transcription factors, expression of specific cell-surface proteins, production of ECM-degrading enzymes, and expression of specific micro RNA (Kalluri & Weinberg, 2009). Many secreted polypeptides are implicated in the EMT program, and their corresponding intracellular transduction pathways form highly interconnected networks. EMT is triggered by an interplay of extracellular signals, including components of the extracellular matrix, soluble growth factors, **such as members of the TGFβ and FGF families, EGF and HGF / SF**. The activation of signalling pathways Wnt, Notch, and BMP4 also results in the activation of transcriptional regulators SNAI 1 and SNAI 2, which regulate the changes in gene-expression patterns that underlie EMT. These complex networks orchestrate the disassembly of junctional complexes and the changes in cytoskeletal organization that occur during EMT (Moustakas & Heldin, 2007; Thiery & Sleeman, 2006).

TGF-β signaling plays a critical role in EMT either through downstream Smad signaling or through Smad-independent pathways (Xu et al., 2009). Transcription factors Snai 1 and Snai 2 are repressors of E-cadherin gene expression and are considered as key regulators of EMT and function overall as an epithelial phenotype repressor (Savagner, 2001). In cell culture, the experimental overexpression of Snai 2 or Snai 1 is sufficient to induce epithelial cells to undergo EMT (Batlle et al., 2000; Cano et al., 2000; Bolós et al., 2003). Leroy and Mostov (2007) suggested that a primary function of Snai 2 during EMT is promoting epithelial cell survival.

Numerous pathways have been described *in vitro* that control phenotype transition in specific cell models. *In vivo* developmental studies suggest that transcriptional control, activated by a specific pathway involving Ras, Src and potentially the Wnt pathway, is an essential step. Molecular hallmarks for EMT include increased expression of N-cadherin and vimentin, nuclear localization of β-catenin, and increased production of the transcription factors, such as Snai 1, Snai 2, Twist, ZEB1, and ZEB2, that inhibit E-cadherin production. Small, non-coding microRNA molecules regulate gene expression by interacting with multiple mRNAs. The miR-200 family of microRNAs inhibits EMT transition by directly targeting E-cadherin transcriptional repressors ZEB1 and ZEB2. The loss of expression of the miR-200 is of critical importance in the repression of E-cadherin by ZEB1 and ZEB2 during EMT, thereby enhancing migration and invasion during cancer progression (Korpal et al., 2008). It was demonstrated that ZEB1, a crucial inducer of EMT in various human tumours, triggers a microRNA-mediated feedforward loop that stabilizes EMT and promotes the invasion of cancer cells (Burk et al., 2008). Using the epithelial Madin–Darby canine kidney (MDCK) cell line model, Gregory and colleagues demonstrated the existence of an autocrine TGF-β/ZEB/miR-200 signaling network that regulates cell plasticity and transition between epithelial and mesenchymal states (Gregory et al., 2011).

Many signaling pathways orchestrate the concerted regulation of an elaborate gene program and a complex protein network, needed for the establishment of new mesenchymal-like phenotypes after disassembly of the main elements of epithelial architecture, such as desmosomes, tight adherens and gap junctions. Loss of epithelial marker E-cadherin is an early event that preceds other cell alterations during EMT. During EMT, adherens, junctions and desmosomes are at least partially dissociated; a profound cytoskeleton reorganization takes place, involving switch expression from keratin- to vimentin-type intermediate filaments.

Hepatocyte Growth Factor

Hepatocyte growth factor (HGF), a large multifunctional polypeptide, is known to play a central role during embryonic development, wound healing, regeneration, and tumorigenesis. Originally, HGF was isolated for its ability to induce hepatocyte growth *in vitro* (Nakamura et al., 1987) and was independently isolated as a scatter factor (SF), which stimulated cell dissociation and motility *in vitro* (Stoker et al., 1987). HGF/SF is secreted by mesenchymal cells and induces a variety of tissue-specific morphogenic programs in epithelial cells. HGF, upon binding to its cell surface receptor c-Met, exerts its pleiotropic effects on cell growth, cell motility, and morphogenesis in a wide range of cellular targets including epithelial and endothelial cells, hematopoietic cells, neurons, and hepatocytes. SF/HGF, as well as its receptor c-Met, is highly expressed in developing tissues during embryogenesis, and both are essential for the embryonic development (Sonnenberg et al., 1993). The subsequent tyrosine phosphorylation of c-Met initiates a complex cascade that involves the recruitment of several adaptor proteins and the activation of a great many of signal transducers that include the extracellular signal-regulated kinase (ERK/MAP); phosphatidyl-inositol-3 kinase (PI-3 K); tyrosine phosphatase SHP2; phospholipase C-γ; Src; Shp-2 phosphatase; STAT3 and the adapter proteins Crk, Grb2, She, and Gabl (Weidner et al., 1998; Birchmeier et al., 2003; Zhang & Vande Woude, 2003). Liang and Chen (2001) found that HGF/SF selectively increases the expression of integrin α_2 in MDCK cells. Then, the authors demonstrated that the level of integrin α_2 expression induced by HGF/SF correlated with the extent of cell scattering and activation of ERK in the cells. The results suggest that the increased expression of integrin α_2 by HGF is at least partially required for cell scattering, and the duration of ERK activation is likely to be a crucial determinant for cells to activate integrin α_2 expression and cell scattering. However, increased expression of integrin α_2 alone is not sufficient for scattering MDCK cells. HGF/SF takes part in different forms of morphogenetic events. For example, HGF/SF alters the polarity of MDCK cell monolayers (Balkovetz et al., 1997) and can regulate tubule formationt during early development of the metanephros (Woolf et al., 1995), in cancer development (Elliott et al., 2002) and wound healing (Chmielowiec et al., 2007). HGF/SF also initiates a program of cell dissociation and increased cell motility that promotes cellular invasion through extracellular matrices, closely resembling tumor metastasis *in vivo* (Zhang & Vander Woude, 2003). It was found that the over-expression of HGF/SF is associated with a high risk for metastasis and recurrence in patients with papillary thyroid carcinoma (Ramirez et al., 2000).

Two naturally occurring truncated forms of HGF/SF (HGF/NK1 and HGF/NK2) retain a high-affinity binding to the c-Met receptor. While HGF/NK1 stimulates tubulogenesis by both epithelial and endothelial cells, HGF/NK2 antagonized the morphogenic effect of full-length HGF/SF (Montesano et al., 1998). It is thought that HGF/SF can induce morphogenic processes in diverse epithelial cells, but exact morphogenic events are determined by intrinsic programs of these epithelia (Brinkmann et al., 1995).

Transforming Growth Factor β

TGF β signaling is one of the major regulatory networks in EMT (Akhurst & Derinck, 2001; Wakefield & Roberts, 2002; Thiery & Sleeman, 2006; Talbot et al., 2012; Yang et al., 2000). It exerts potent control over cell proliferation, differentiation, apoptosis, adhesion, invasion, and interactions with the cellular microenvironment.The induction of EMT by TGF β was first recognized in cell culture. Upon TGF β treatment, mammary epithelial cells changed from cuboidal to an elongated spindle shape, and showed decreased expression of epithelial markers, such as E-cadherin, and enhanced expression of mesenchymal markers fibronectin and vimentin (Miettinen et al., 1994). These changes were accompanied by increased motility. Consistent with their binding to the same receptor complexes, TGF β1, TGF β2 and TGF β3 share the capacity to induce EMT in epithelial cells. However, Foitzik and colleagues (1999) found thatonly TGF-β2 isoform is the inducer of hair follicle morphogenesis in mice. Similarly to various growth factors that act through tyrosine kinase receptors, TGF β has been shown to rapidly activate PI3 kinase, leading to an activation of the Akt kinase, in diverse cell systems. During EMT, TGF β-induced activation of the receptor complex leads to activation of Smad2 and Smad3 through direct C-terminal phosphorylation. Phosphorylated Smad2 and Smad3 then form trimers with Smad4, and translocate into the nucleus, where they associate and cooperate with DNA-binding transcription factors to activate or repress target gene transcription (Xu et al., 2009). It was found that integrin-linked kinase (ILK) is required for EMT induced by TGF β. It was shown that the PI3K-ILK-Akt pathway, independent of the TGF-β -induced Smad pathway, is required for TGF- β -mediated EMT (Lee et al., 2004). TGF-β1 and Ras co-operate to induce EMT in human keratinocytes by mechanisms that involve MAPKs, Smad2/3 and AP-1 (Davies et al., 2005). Double immunohistochemistry staining showed that in normal rat kidney tubular epithelial cell line NRK52E, the addition of TGF-β1 to confluent cell cultures resulted in a loss of the epithelial marker E-cadherin and *de novo* expression of alpha-SMA (Fan et al., 1999).

The importance of the p63 gene expression was also demonstrated in the EMT process. p63 is transcribed from two distinct promoters, resulting in Tap63 and ΔNp63 isoforms based on alternative transcription start sites. Tap63 is highly expressed in basal cell nuclei of stratified epithelia in adult skin and in early-stage skin morphogenesis and is responsible for initiation of epithelila stratification (Koster et al., 2004). Oh et al. (2011) demonstrated that ΔNp63α–transduced normal human epidermal keratinocytes in the presence of intact TGF β signaling underwent morphologial changes resembling mesenchymal cells and acquired the EMT phenotype in senescent cultures. These cells gained the ability to be differentiated to osteo/odontogenic and adipogenic pathways, resembling mesemchymal stem cells. Treatment with TGF β stimulated EMT in pre-senescent ΔNp63α-transduced cells, while the inhibition of TGF β signaling reversed the EMT phenotype. EMT of normal human epidermal keratinocytes led to increased telomeric length, independent of telomerase activity. Furthermore, these cells expressed enhanced levels of Nanog and Lin28, which are transcription factors associated with pluripotency.

Vimentin

Vimentin is a type III intermediate protein filament, which is the major cytoskeletal component of and expressed in mesenchymal cells. Vimentin is often used as a marker of mesenchymally-derived cells or cells undergoing EMT. During embryogenesis, vimentin synthesis occurs for the first time in the primitive streak stage and is restricted to the primary mesenchymal cells, which cease to produce cytokeratins and desmoplakin (Franke et al., 1982). Some publications demonstrated expression of vimentin in cultured epithelial cells. Human fetal keratinocytes grown on plastic in keratinocyte growth medium express some of the components of the differentiated epidermis, such as involucrin and the high molecular weight keratins. However, these keratinocytes co-express keratins and vimentin and do not form a structured basement membrane (Haake & Lane, 1989). It was shown that both keratin and vimentin are expressed in the epithelial cell line MDCK (Zuk et al., 1989). In cultured keratinocytes derived from normal human skin and grown on plastic in a serum-free defined medium (MCDB 153, with 0.1 mM Ca^{2+} concentration), vimentin expression was demonstrated in the cell cytoplasm by immunohistochemistry with different monoclonal antibodies to vimentin (Richard et al., 1990). In foreskin keratinocytes cultured in a low Ca^{2+} concentration and then switched to a high Ca^{2+} concentration (1.6mM) to induce differentiation, vimentin was present in 40-70% of the basal cells 16 hours after the switch (Van Muijen et al., 1987). Johnson and Spandau (1997) demonstrated that normal human keratinocytes grown *in vitro* co-express both cytokeratins and vimentin. During wound healing *in vivo,* keratinocytes also express both vimentin and cytokeratins. In human mammary epithelial MCF10A cells, vimentin was transiently expressed in migrating cells in an *in vitro* wound-healing model and disappeared when the cells became stationary after wound closure. Vimentin was functionally involved in the migratory status of epithelial cells: transfection of the antisense vimentin cDNA into MCF10A cells significantly reduced both their ability to express vimentin and cell migratory speed (Gilles et al., 1999). This data suggests that cultured cells of epithelial origin are in a state of partial EMT.

Fibronectin

Fibronectin is a high-molecular weight glycoprotein of the extracellular matrix that binds to integrins. Fibronectin plays a major role in cell adhesion and migration and is important for wound healing and embryonic development. Insoluble cellular fibronectin is a major component of the extracellular matrix. It is secreted by various cells, primarily fibroblasts, as a soluble protein dimer and is then assembled into an insoluble matrrix in a complex cell-mediated process. It is thought that fibronectin plays a central role in wound healing due to an augmentation of cell migration. The deposition of fibronectin underlying the tubule-lining epithelium serves to enhance cell proliferation and migration, and facilitates the branching tubulogenesis of MDCK cells (Jiang et al., 2000).

Fibronectin was not detected in a normal differentiating epidermis, but it was detected in or near the epidermal basement membrane. On the contrary, cultured epidermal keratinocytes actively produce fibronectin. The synthesis and distribution of fibronectin from newborn rat epidermal cells cultured in the absence of a feeder layer was shown by an indirect immunofluorescence staining of cultures. The localizations of fibronectin and its close association with actin also suggest that it is involved in keratinocyte adhesion and is related to the internal cytoskeletal organization of these cells (Gibson et al., 1983). Human keratinocytes cultured in a serum-free, low-calcium medium without a fibroblast feeder layer (conditions that prevent terminal differentiation) can synthesize, secrete, and deposit fibronectin in the extracellular matrix (Kubo et al., 1984; O'Keefe et al., 1984). In cultured human epidermal keratinocytes derived from foreskin only a proportion of the cells produced fibronectin that correlated with the level of differentiation of keratinocytes (Brown & Parkinson, 1983).

E-Cadherin

Classical cadherins are the transmembrane components of the adherens junction. Cadherins mediate cell-cell adhesion through their extracellular domains, and their cytoplasmic domains bind tightly to β-catenin, a cytoplasmic protein that interacts with α-catenin, which in turn anchors to the actin cytoskeleton, either directly or indirectly via actin-binding proteins α-actinin and vinculin (Niessen & Gottardi, 2008).

Epithelial cells typically express E-cadherin, whereas mesenchymal cells express various cadherins, including N-cadherin, R-cadherin and cadherin-11. The loss of E-cadherin expression plays a central role in the EMT program. Downregulation of E-cadherin and upregulation of N-cadherin during EMT is reminiscent of the cadherin switching that is seen during normal embryonic development. The cadherin molecules at adheren junctions have multiple isoforms. Cadherin isoform switching occurs during normal developmental processes to allow cell types to segregate from one another (Wheelock et al., 2008). In a classical model system, TGF β1-mediated EMT in mammary epithelial cells, it was found that the morphological changes associated with EMT preceded the downregulation of E-cadherin. It was concluded, that cadherin switching is necessary for increased motility but is not required for the morphological changes that accompany EMT (Maeda et al., 2005). N-cadherin promotes motility when expressed by epithelial cells (De Wever et al., 2004). Interestingly, cells that express significant amounts of E-cadherin and only a small amount of N-cadherin still have increased motility. This suggests that N-cadherin plays an active role in cell motility that E-cadherin cannot suppress (Nieman et al., 1999). Transcriptional repression of the E-cadherin gene induces cellular responses leading to the conversion of an epithelial phenotype into an invasive mesenchymal phenotype.

Transcriptional repression of the E-cadherin gene induces cellular responses leading to the conversion of epithelial cells into invasive mesenchymal cells. Loss of epithelial marker E-cadherin is an early event that preceds other cell alterations during EMT. PI-3K/AKT pathways are involved in the regulation of E-cadherin, cell motility and invasion. PI 3-K and Grb2, together with Ras and MAPK, provide intracellular pathways, which may lead to mitogenesis, cell scattering or morphogenesis. Then, intracellular effectors are recruited and activated, leading to the integration of the initiatory signal into a specific phenotype (Stuart et al., 2000).

It was shown that the transient expression of exogenous E-cadherin in fibroblastoid cells reduced cell growth **and decreased** β-catenin activity, controlled by E-cadherin in a cell-adhesion independent manner (Stockinger et al., 2001). Many kinds of tumors are characterized by the lack of E-cadherin expression (Moll et al., 1993). Interestingly, transfection of either metastatic cells or normal embryonic fibroblasts with the E-cadherin gene converts them to the epithelial phenotype (Hay, 1995). Embryonic mesenchymal cells aggregate when transfected with the E-cadherin gene. They stratify to form a desmosome-rich epithelium that express keratin and desmoplakin when cells are grown on glass or collagen (Vanderburg & Hay, 1992).

Extracellular Matrix

Normal epithelial cells grown on top of collagenous matrices form epithelial sheets, whereas mesenchymal cells invade underlying ECM. A number of adult and embryonic epithelia, when suspended within native type I collagen gels, give rise to elongated mesenchyme-like cells that migrate freely within the 3D matrix. The morphology of these cells is indistinguishable from the true mesenchymal cells at light and ultrastructural levels. Greenburg and Hay (1986) derived cells of lens epithelium from 12-day old chick embryos. When suspended within collagen gels, cells lost epithelial morphology, phenotype, and cytodifferentiation and formed mesenchyme-like cells that lacked the ability to synthesize lens-specific delta-crystallin protein, type IV collagen, and laminin. However, they expressed type I collagen, a characteristic of mesenchymal cells. These tissue phenotype changes, which occur during the EMT, were stable under the conditions studied. When mesenchyme-like cells were removed from the gel and replated onto two-dimensional surfaces, they remained bipolar, freely invaded collagen matrices, and were unable to synthesize delta-crystallin protein (Greenburg & Hay, 1986). MDCK explants cultured on type I collagen gel give rise to isolated fusiform-shaped cells that show mesenchymal cell polarity and migrate over the gel surface. These cells extend pseudopodia and filopodia, lose cell membrane specializations, and develop an actin cortex around the entire cell. Unlike true mesenchymal cells, fusiform cells display an intermediate filament phenotype: they produce both keratin and vimentin, continue to express laminin, do not turn on type I collagen, and do not invade the gel. Fusiform cells are not apically-basally polarized, but show mesenchymal cell polarity (Zuk et al., 1989). However, MDCK cells suspended alone in collagen gels formed spherical cysts, whereas they form branching tubules in the presence of fibroblasts (Montesano, 1991a, b). Some studies described EMT as a result of cell-matrix interactions, particulary via the influence of collagen types I and III expressions on the E-cadherin gene expression and disruption of the E-cadherin adhesion complex in epithelial tumors. The molecular mechanisms involved in collagen-induced EMT include activation of integrins, activation and translocation of the focal adhesion kinase to the E-cadherin/catenin complex as well as inhibition of the phosphatase PTEN (Imamichi & Menke, 2007). Increased matrix metalloproteinase expression may stimulate fibrosis, tumorigenesis, and tumor progression by inducing a specialized EMT in which epithelial cells transdifferentiate into activated myofibroblasts (Radisky et al., 2007).

Collective Migration of Epithelial Cells

Epithelial cells can migrate as single free entities and also as continuous sheets. When placed in culture, epithelial cells dramatically change their phenotype. Single epithelial cells protrude lamellipodia and acquire fibroblastoid morphology typical of migrating cells (Greenberg & Hay, 1982). Collective cell migration is a special characteristic of the motility behavior of epithelial cells *in vivo* and *in vitro*. It has been suggested that epithelium will **not tolerate a free edge (Trinkaus, 1984). Disruption of an epithelium's** integrity through wounding initiates a migration of surrounding cells to close up the gap. When migrating as a coherent sheet,epithelial cells maintain their cell-cell contacts. Scratch wound closure in confluent epithelial cultures resembles cell-sheet movement during re-epithelization in wound healing *in vivo*. Parallels between the repair of epithelial tissues and embryo morphogenesis were also suggested (Martin &Parkhurst, 2004).

MDCK is the most extensively studied cell line in epithelial cell biology. MDCK cells can move as dispersing individual cells following treatment of subconfluent cultures with HGF/SF or as continuous sheets after mechanical disruption of a confluent monolayer. In the moving cell sheet, the marginal cells of the leading edge protrude typical lamellipodia, and submarginal cell protrusions also displayed lamellipodia with charateristic morphology and dynamics (Farooqui & Fenteany, 2005), although the lamellipodia of marginal cells were longer and broader. It seems that submarginal cells individually generate their own motile force by a protrusive crawling mechanism and are not just being pulled by cells in front of them. In the wound, some transient cell proliferation occurs at a delay after wounding but the complete pharmacological blockage of proliferation does not affect the rate of wound closure. After, the wound re-epithelization spread cells revert to their normal cuboidal morphology and the original cell density of the unbroken sheet is restored (Farooqui, Fenteany, 2005). Closing cell monolayer wounds involves Rac-, phosphoinositide- and c-Jun-N-terminal kinase-dependent migration (Altan & Fenteany, 2004). Poujade et al. (2007) studied a situation when the free surface was presented to the confluent MDCK cell layer with no damage made to the cells. They found that a sudden release of the available surface is sufficient to trigger a collective migration of cells, which was independent of cell proliferation that mainly took place on the fraction of the surface initially covered with cells. These authors observed both extremely complex and coordinated long-range motility within the epithelium, characterized by a duality between collective and individual behaviors of cells. On the one hand, the velocity fields within the monolayer are of a very long range and involve many cells in a coordinated way. On the other hand, very active leader cells have nonepithelial, fibroblast-like morphology and behavior. These leader cells preceded a small cohort of cells and destabilized the border by a fingering instability. However, in the presence of exogenous HGF, the migrating cell sheet had a higher average velocity of the border and no leader cells. In our laboratory, the migration of single colonies formed by human epidermal keratinocytes was studied. On a plastic substrate, epidermal cells form multicellular colonies manifesting collective cell migration. Transmission electron microscopy showed that elongated leading edge cells developed extensive lamellipodia, and no less than four layers of submarginal cells also produced lamellipodia in the direction of colony migration.

Migrating keratinocytes contacted the substrate with only a minor part of their surface. A part of the cell surface attached to the substrate usually formed long lamellipodium in the direction of the colony migration and a short one in the opposite direction, which contacted the long lamellipodium of the following cell. Furthermore, migrating cells produced short processes contacting the substrate and were interconnected with neighbor cells through rare desmosomes and short processes (Figure 1). On the contrary, epidermal cells in the middle of the colony retain more cuboid morphology, which is typical for confluent monolayer cells (Vorotelyak et al., 1996).

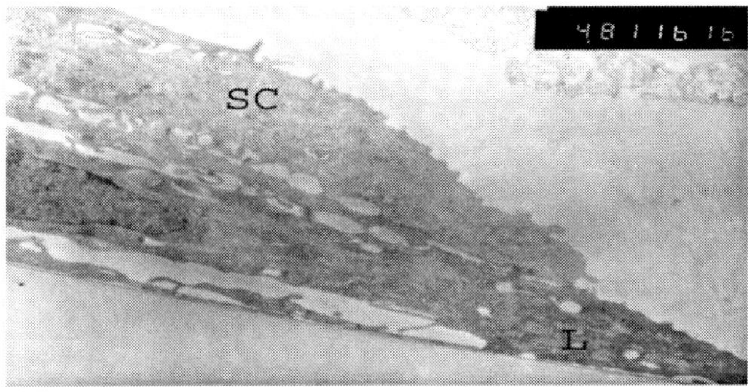

Figure 1a. Transmission electron microscopy of the leading cell of migrating keratinocyte colony forming long lamellipodium (L) in the direction of migration and suprabasal cell (SC) crawling over the leading cell. Note small outgrowth of the leading cell in the opposite direction and several remaining desmosomes.

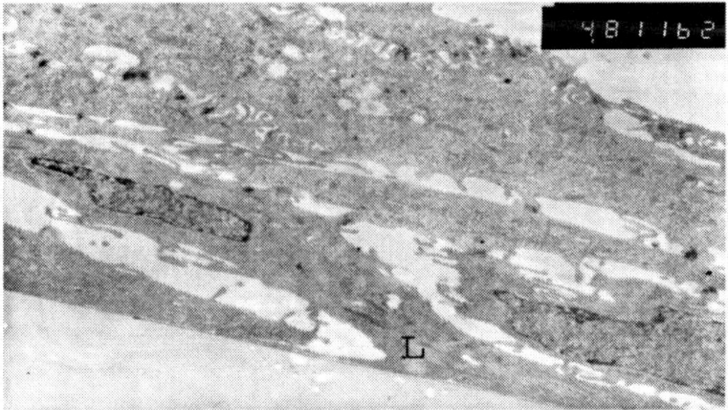

Figure 1b. Transmission electron microscopy of the migrating submarginal cells of the colony. Two cells of the basal layer form long lamellipodia (L) contacting with substrate in the direction of movement. They also form multiple outgrowths. Desmosomes are seen.

Tubulogenesis

Formation of highly organized networks of tubules, known as tubulogenesis, is a developmental process common to the formation of many organs, such as the lung, salivary gland, mammary gland, pancreas, mammalian kidney, vasculature, and *Drosophila* trachea. Tubulogenesis, the most complex process of epithelal tissue organization, is regulated by a wide range of growth factors and their receptors, transcription factors and cell–cell and cell–extracellular matrix (ECM) adhesion molecules.

To build a structure, a population of epithelial cells must coordinate many behaviors: proliferation, migration, adhesion, polarization, differentiation and death.

The classic *in vitro* assay of tubulogenesis was made from the development of an *in vitro* model of tubulogenesis using MDCK epithelial cells, which have properties of the kidney distal tubule and collecting duct. Montesano et al. (1991a, b) demonstrated that MDCK cells suspended alone in 3D collagen gels form spherical cysts, which produced long cytoplasmic processes extending from the cyst wall into the surrounding collagen matrix. In the presence of fibroblasts the cysts produced radial outgrowths of branching epithelial cords from the cyst wall. Then, it was shown that MDCK cells grown as a monolayer on fibroblast-containing collagen gels invade the underlying matrix, within which they form a network of tubules. The authors also found that a fibroblast-conditioned medium mimics the effects of fibroblasts by producing tubulogenesis by MDCK cells. Then the involvement of paracrine HGF in morphogenetic epithelial-mesenchymal interactions was demonstrated (Montesano et al., 1991a,b). The *in vitro* branching tubulogenesis is reminiscent of the structures found in many epithelial organs; thus, cysts and tubules can be considered as the basic building blocks from which more complex epithelial organs are formed.

In cultured MDCK cells, Pollack et al. (1998) examined the dynamics of tubule formation and desribed a multistep model of tubulogenesis. During the **first stage,** the first morphological event is the elongation of cells in the apicobasal dimension and protrusions into the surrounding collagen gel while remaining attached to the cyst. Apical and basolateral membranes are morphologically altered, but cell polarity and cell junctions are maintained.

The second stage of tubulogenesis, chain formation, is characterized by single-file chains of cells that form either linear or branching structures. In the chains cells lose apico-basal polarity but retain cell-cell contacts. The apical membrane protein gp135 and the basolateral protein desmoplakin are lost from the plasma membrane and appear in the cytoplasm. The adherens junction protein E-cadherin redistributes uniformly throughout the plasma membrane of cells in the chain. At the next stage, called the cord, chains of cells thicken into cords, which are two to three cells in diameter and are devoid of lumina. Apical polarity is again observed, while E-cadherin remains randomly distributed. Then, a maturation of cords into tubules proceeds and discontinuous lumina develop. The lumina enlarge and become continuous, resulting in tubules. Apical and basolateral membranes of cells of the tubule become clearly polarized and the arrangement of cell junctions in polarized epithelial cells is completely restored. The basolateral polarity of E-cadherin in lumen-containing regions of developing tubules is restored. Thus, tubulogenesis develops via a *de novo* formation of lumina in MDCK cells.

Similar events were found in the process of tubulogenesis of human postnatal epidermal keratinocytes in collagen gel; however, no clear lumina developed in the mature cords (Shinin et al., 2002). Tubulogenesis in human keratinocytes cultured in collagen gel was promoted by EGF, KGF, and HGF (Gnedeva et al., 2009). The tips of tubules consisted of small cells organized in tightly condensed patches. The inside of tubules contained few cells interspersed with lumina (Chermnykh et al., 2010). Unlike tubulogenesis in kidney or lung epithelia, branched tubulogenesis was never detected in epidermal keratinocytes.

An immunohystochemical analysis of cytokeratins expression demonstrated that differentiated, as well as undifferentiated, keratinocytes comprised tubules. The pattern of cell distribution could differ on sections. In many instances, K14-positive cells were localized in peripheral layers of tubules, while inner parts contained few nucleated elements. Confocal microscopy proved an active involvement of K14-positive cells in tubulogenesis. Multiple K14-positive cells were found in the growing tip of a tubule. Keratin 10-positive cells localize in tubules without regular patterns, although they are rarely found in the outmost cell layer. We suggest that in growing, tubules must first migrate low-differentiated cells. While moving through the gel they differentiate and slow down, and are used by newcomer cells as a substrate for migration. The first manifestation of EMT in epidermal keratinocytes is the formation of long cytoplasmic extensions into the matrix. In this model, the formation of cell extensions and single-file chains of cells was independent of cell proliferation. Proliferation was of importance at the cord stage, and an inhibitor of DNA synthesis, mitomycin C, suppressed the formation of cord structures. These results suggest the partial dissociation of cell migration and cell proliferation during tubulogenesis by human epidermal keratinocytes (Shinin et al., 2002).

In MDCK cells (Pollack et al., 1998) and in human epidermal keratinocytes (Shinin et al., 2002), tubulogenesis occurs only in cysts or big cell clusters embedded in collagen gel, probably as a result of intercellular morphogenetic interactions. Individual cells scattered in collagen gel do not contribute to the tubule development. In human keratinocytes, as well as in MDCK cells, cell–cell contacts were retained at all stages of tubulogenesis.

In addition, the composition of the surrounding ECM has been shown to modulate HGF/SF-induced morphogenesis. Tubulogenesis was completely abrogated in Matrigel. It was found that early responses to SF/HGF were matrix-independent. Changes included increased paracellular spacing between normally closely apposed lateral membranes, and the formation of filopodial processes, indicating a partial motile response. Cell–cell contacts were maintained, with the persistence of cell junctions. Therefore, while one or a number of ECM components are preventing HGF/SF-primed cells from undergoing an invasive and/or migratory program, non-permissive matrices do not prevent HGF/SF signaling to the cell. Later matrix-dependent responses, which occurred in type I collagen but not Matrigel, included the formation of basal protrusions that comprise two or more neighbor cells, which extend to form nascent tubules. A modified polarity of cells comprising the basal protrusions was evident, with a marker for the apical membrane being found in the same region as adherens junctions and desmosomes, typically localized at lateral membranes. A model for HGF/SF-induced tubulogenesis was proposed in which tubules form from basal protrusions of adjacent cells. This mechanism of *in vitro* tubule formation has many similarities to the reported *in vivo* epithelial tubulogenesis (Williams & Clark, 2003).

Downstream effectors of HGF/SF regulate successive stages of tubulogenesis. Activation of ERK is necessary and sufficient for the initial stage, during which cells depolarize and migrate. ERK becomes dispensable for the latter stage, during which cells repolarize and differentiate. For the late stage the activity of matrix metalloproteases (MMPs) is essential, but not during the initial stage. Thus, Erk and MMPs define two programs that act in sequence (O'Brien et al., 2004).

In vivo, integrin-linked kinase (ILK) was markedly induced in renal tubular epithelia in mouse models of chronic renal diseases, and such induction was spatially and temporally correlated with tubular EMT. Moreover, inhibition of ILK expression by HGF was associated with blockades of tubular EMT and an attenuation of renal fibrosis. These findings suggest that ILK is a critical mediator for tubular EMT and likely plays a crucial role in the pathogenesis of chronic renal fibrosis (Li et al., 2003).

Hair Follicle

Tremendous progress in understanding HF development and cycling provides insights into the key questions in morphogenesis. Now, it is known that some of the major events of development, such as cell proliferation and differentiation, cell signaling and cell rearrangements, are involved and precisely regulated in HF renewal. Reciprocal signaling between epithelial cells of the hair bulb and mesenchymal cells of the dermal papilla (DP) reach their maximum during late anagen (Botchkarev & Kishimoto, 2003; Rendl et al., 2005). It is of interest that almost all signaling pathways, including Shh, Wnt, Bmp, FGF, TGF, and Notch, involved in embryogenesis, are involved in HF renewal. The morphogenetic processes that take place during hair development are recapitulated in adult follicles during hair cycling.

During hair follicle development, epithelial-mesenchymal cross talk between epidermal keratinocytes and DP cells plays a pivotal role in producing a mini-organ that shows cyclic activity during postnatal life with periods of active growth, involution and quiescence. When anagen growth is activated, a new lower follicle must be regenerated from the adult stem cells. The follicular DP comprises a group of mesenchymal cells at the base of the hair follicle, which has a crucial role in hair follicle development and regulates the postnatal hair growth cycle (Cotsarelis, 1990; Yang & Cotsarelis, 2010). DP cells have a distinct gene expression signature compared with non-DP dermal fibroblasts (Rendl et al., 2008). The matrix is a proliferative zone surrounding the DP. Proliferating matrix keratinocytes contribute to hair follicle cycling and the formation of the hair shaft. Shimaoka et al. (1994; 1995) studied the effect of HGF on mouse and human hair growth in organ and cell culture systems, and Jindo et al. (1994) investigated the effect of HGF on vibrissal hair follicles isolated from newborn mice in culture. It was found that DP cells expressed HGF and stimulated hair growth, DNA synthesis, and mitogenic activity in hair bulb-derived keratinocytes.

Hirai and co-workers examined the expression of E- and P-cadherin in the histogenesis of the epidermis and hair follicles using the upper lip skin of the mouse. They found that P-cadherin was expressed exclusively in the proliferating region of the epidermis and hair follicle: in the germinative layer of the epidermis, the outer root sheath and the hair matrix. E-cadherin was coexpressed in these regions but was also detected in non-proliferating regions such as the intermediate layer of the surface epidermis and the immature regions of the inner root sheath. It was suggested that cadherins present in epidermal cells are involved not only in maintaining the arrangement of these cells, but also in inducing dermal condensation (Hirai et al., 1989).

Budding morphogenesis is a fundamental problem in developmental biology. A simple bud-like structure initiates the formation of many organs, including lungs, spinal cord, mammary glands, and hair follicles (Hogan, 1999). During the early phase of hair follicle morphogenesis within the developing skin, epidermal keratinocytes reorient to form an epithelial bud structure by changing their polarity and cell–cell contacts via switching from E-cadherin to P-cadherin-based adhesion junctions. At the mesenchymal–epithelial interface, the cadherin switch persisted throughout follicle development. In adult hair follicles, as in buds, P-cadherin was expressed only in the cell layers near the DP, whereas E-cadherin was expressed in the more external layers when cell differentiation begins. This transition from E-cadherin to P cadherin synthesis coincides with the increase in cell migration and activation of cell proliferation. Then, it was demonstrated that transgenical elevation of E-cadherin molecules in mouse skin blocked hair follicle morphogenesis, suggesting that E-cadherin down-regulation is a critical event in budding morphogenesis (Jamora et al., 2003). It was found that TGF $\beta2$ signaling is necessary to transiently induce the transcription factor Snai 1 and activate the Ras-mitogen-activated protein kinase (MAPK) pathway in the bud. Snai 1 expression was detected in the hair bud of E17.5 skin, but not in later stages of development (Jamora et al., 2005). Slug (Snai 2) was prominently expressed in placodes and hair pegs of the developing hair follicle on E14 **(Parent et al., 2010)**. TGF β 1 **induction** of transcription factors Snai 1 and Snai 2 may be considered as key regulators of EMT in epithelial cell lines (Bolós et al., 2003; Peinado et al., 2003). Keratinocytes at wound margins undergo partial EMT and Snai 2 is an important modulator of successful wound repair in adult tissue and may be critical for maintaining epidermal integrity in response to chronic injury (Hudson et al., 2009).

It is possible to link HGF/SF production, by dermal papilla and downregulation of E-cadherin, as a typical event of EMT program. Development of a hair follicle begins via budding, when cells extend out from the epithelium in a direction orthogonal to the epithelial plane. Later, this outgrowth forms a peg, which is reminiscent of a cord produced during tubulogenesis in MDCK cells (Pollack et al., 1998). HGF/SF is transiently localized in the DP during morphogenesis and during anagen III-to-catagen II of the hair cycle.

Epithelial cells in the vicinity of HGF/SF-producing DP fibroblasts express a Met receptor. In addition, HGF/SF overexpression retards catagen and hair follicle regression *in vivo*. This data provides suggestive evidence for an involvement of this growth factor-signaling system in hair growth control, both during morphogenesis and the cyclic remodeling of a hair follicle (Lindner et al., 2000).

Fujie et al. (2001) demonstrated that human DP cells stimulated the chemotactic migration of human hair outer-root sheath cells. They found that keratinocyte migration was dramatically increased toward IGF-I and HGF, however neither VEGF nor TGF-β1 showed any effect on the chemotaxis. In the three-dimensional model of a living skin equivalent, with keratinocytes grown on top of collagen gel, DP cells, embedded into the gel, stimulated the outgrowth of keratinocytes into a collagen matrix and the formation of protrusions. In a modified model of tubulogenesis, human keratinocytes were tightly packed in wells within collagen gel.

In this model, DP cells initiated the formation of tubules, which reached 1–2 mm length, on average, and had variable shapes with different length-to-width ratios (Figure 2).

The tips of tubules consisted of small cells organized in tightly condensed patches. Multiple K14-positive cells were found in the growing tips of tubules, the inside of tubules contained few cells interspersed with lumina. (Chermnykh et al., 2010). The length of the tubules initiated by DP cells *in vitro* is comparable with the length of anagen hair follicles and, presumably, is genetically regulated. It is of interest that the growth of epithelial tubes during the development of the Drosophila tracheal system was demonstrated to be under genetic control.

The diameters and lengths of tubes are regulated independently: tube length increases gradually throughout development, whereas tube diameter increases abruptly at discrete moments in development. In the embryonic and larval respiratory system of Drosophila, a genetic analysis showed that tube sizes are specified early by branch identity genes, and the subsequent enlargement of branches to their mature sizes and maintenance of the expanded tubes involves a new set of tube expansion genes (Beitel & Krasnow, 2000).

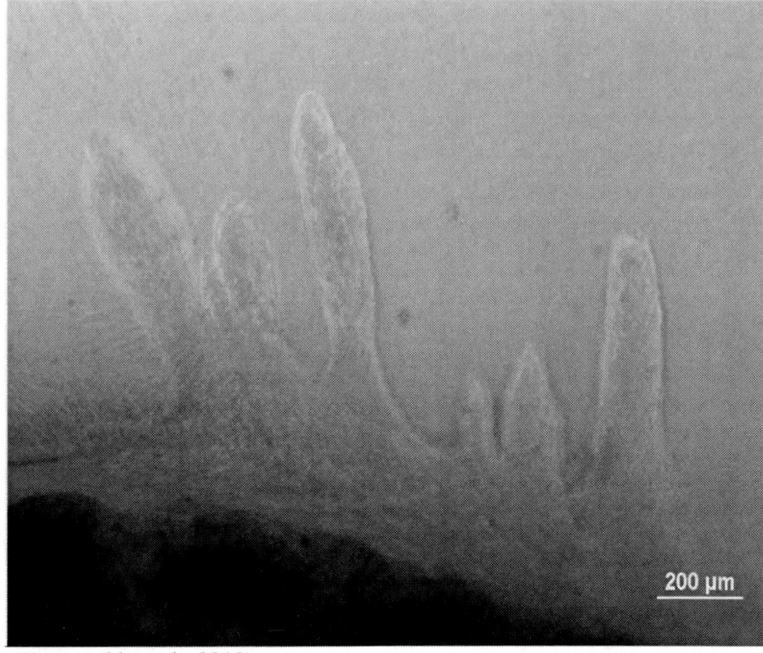

(from Chermnykh et al., 2010).

Figure 2. Tubulogenesis from a condensed aggregate of keratinocytes in collagen gel induced by dermal papilla cells.

EMT and Carcinogenesis

The EMT at least superficially resembles the metastatic transition of normal cells during epithelial tumor progression. Essential for embryonic development, EMT is nevertheless potentially destructive if deregulated, and it is becoming clear that the inappropriate utilization of EMT mechanisms is an integral component of the progression of many tumors of epithelial tissues.

An increasingly accepted concept is that EMT, which is a normal developmental program, is an important preliminary step in tumor progression and metastasis (Thompson et al., 2005; Yang & Weinberg, 2008; Ding et al., 2010; Hugo et al., 2007). There is growing evidence suggesting that the alterations in tumor tissue architecture take place through the EMT. The essential features of EMT are the disruption of intercellular contacts and the enhancement of cell motility, thereby leading to the release of cells from the parent epithelial tissue.

The resulting mesenchymal-like phenotype is suitable for migration and, thus, for tumor invasion and metastatic progression. Although the molecular bases of EMT have not been completely elucidated, several interconnected transduction pathways and a number of potentially involved signalling molecules have been identified. Most of these pathways converge on the down-regulation of the epithelial molecule E-cadherin, which is somatically inactivated in many diffuse-type cancers (Guarino et al., 2007).

Aberrant wound healing with chronic inflammation can promote malignant transformation, and a wide variety of human cancers, including those arising from the lung, liver, pancreas, bone, and skin, are associated with injury. Cancers are often compared with wounds that do not heal (Dvorak, 1986; Riss et al., 2006; Schäfer & Werner, 2008). The association between wounding and tumorigenesis was demonstrated by studying human basal cell carcinoma-like tumors that originate from hair follicular stem cells which wounding recruits from the follicle bulge and secondary hair germ (Arwert et al., 2012; Wong & Reiter, 2011).

The most common epithelial tumors of the skin are basal cell carcinomas, nearly all of them arising from hair follicles, which harbor cutaneous epithelial stem cells. It is now believed that basal-cell carcinomas arise from undifferentiated cells of the outer root sheath. It has been suggested that their formation is similar to the hair follicle development, and conditional skin tumorigenesis recapitulates the hair follicle growth cycle. Temporally and spatially constrained Hedgehog (Hh) signaling regulates cyclic growth of hair follicle epithelium while constitutive Hh signaling drives the development of basal cell carcinomas. This data supports the concept that basal cell carcinomas represent an aberrant form of hair follicle organogenesis (Hutchin et al., 2005).

Morris and her colleagues demonstrated that a slowly cycling subpopulation of adult murine epidermal cells retained radioactive carcinogens and suggested that epidermal slowly cycling cells may be relevant to the initiation phase of two-stage carcinogenesis, and epidermal targets of carcinogenic action were exactly found in those regions of the hair follicle where slowly cycling label-retaining cells have been discovered (Morris et al., 2000). The EMT program may contribute to the formation of metastasis in at least two distinct ways: cells that have undergone EMT seem to possess the mesenchymal traits necessary for dissemination and the stem-cell properties required to generate macroscopic secondary tumors. This may suggest that therapeutic resistance of cancer stem cells may be mediated by cell phenotypes that are conferred by the EMT program.

Stem Cell Markers

Recently, it was discovered that certain epithelial cells that pass through an EMT acquire the self-renewing traits associated with normal tissue stem cells and cancer stem cells. On the other hand, cancer stem cells may express both stem–related traits and EMT attributes. Some data gives the basis to propose that differentiated mammary epithelial cells may pass through an EMT and enter into a mesenchymal state.

Mani et al. (2008) studied EMT in the immortalized human mammary epithelial cells and in freshly sorted, primary mouse mammary epithelial cells and noted that the transient induction in mammary epithelial cells of the EMT yields great increases in their ability to form mammospheres.

The cells from mammospheres acquired a fibroblast-like appearances; downregulated the expression of the mRNA encoding E-cadherin; upregulated mRNAs encoding mesenchymal markers N-cadherin, vimentin, and fibronectin; and acquired many properties of self-renewing stem cells. These stem cell-like cells expressed two cell-surface markers, $CD44^{high}/CD24^{low}$. Their expressions are associated both with human breast cancer stem cells and normal mammary epithelial stem cells as was demonstrated earlier (Mani et al., 2008). It was demonstrated that three signaling pathways, involving transforming growth factor TGF-β **and canonical and noncanonical Wnt** signaling, collaborate to induce the activation of the EMT program and thereafter function in an autocrine fashion to maintain the resulting mesenchymal state.

Downregulation of endogenously synthesized inhibitors of autocrine signals in epithelial cells enables the induction of the EMT program. Conversely, disruption of autocrine signaling by added inhibitors of these pathways inhibits migration and self-renewal in primary mammary epithelial cells and reduces tumorigenicity and metastasis by their transformed derivatives (Scheel et al., 2011).

In another paper (Morel et al., 2008), it was also reported that the induction of EMT in human mammary epithelial cells, following the activation of the Ras-MAPK pathway, produced cells with characteristics of stem cells.

Spheroid cell cultures of the human head and neck squamous cell carcinomas are enriched for cancer stem cell-like cells characterized by a high proportion of ALDH1 positivity, proliferative quiescence, and invasive capacity. The mRNA of the stemness-related genes Sox2, Nanog, and Oct3/4 was significantly increased in these cultures. Spheroid cell cultures had a higher proportion quiescent cells and showed a high-level expression of α-

SMA and vimentin, but a significantly decreased E-Cadherin expression (Chen et al., 2011). It is thought that treatment resistance of cancers to conventional therapy may be connected with the acquisition of hallmarks of EMT by cancer stem cells (Dave et al., 2012).

It is of interest that reprogramming somatic fibroblasts in order to induce pluripotent stem cells by expression of defined factors Oct4, Klf4, c-Myc, and Sox2 was characterized by the induction of MET during the initiation phase of a multistep process (Samavarchi-Tehrani et al., 2010).

Characteristic EMT events take place during the differentiation of human embryonic stem cells in a monolayer culture, including the E-to-N-cadherin switch, increased vimentin expression, up-regulation of E-cadherin repressors Snail and Snai2 and increased activity of metalloproteinases MMP-2 and MMP-9 (Eastham et al., 2007).

Regenerative Medicine

For many years cultivation of the primary epithelial cells in a pure state was unsuccessful because the isolated cells grew very poorly. In 1974, Howard Green and his graduate student James Rheinwald, when culturing teratoma cells, noticed that some cells gave rise to colonies of epithelial appearance against a background of fibroblasts.

The next step was to use a fibroblast feeder layer to support epithelial growth. Fortunately, as a feeder layer, the authors used lethally irradiated cells of the mouse embryo fibroblast line 3T3-J2 that had been obtained in the H. Green laboratory (Todaro & Green, 1963). Due to this approach, rapidly growing colonies were obtained from single epithelial cells (Rheinwald & Green, 1975).

Successive improvements in the cultivation technique made the keratinocyte one of the most proliferative human cell types *in vitro*. Starting with a small skin biopsy, it became possible to produce a number of cultured epidermal cells sufficient to cover the entire body of a human.

Actually, this success was due to paracrine factors secreted by feeder layer fibroblasts. As was demonstrated by Montesano et al. (1991b), single MDCK cells, in the presence of a conditioned medium from the Swiss 3T3 cell line, produced colonies four times greater than in control.

It was also found that many embryo fibroblast lines, such as mouse Swiss 3T3 and BALB 3T3, and human embryonic lung MRC-5 and WI-38 cell lines stimulated EMT in MDCK cells. Moreover, the cultivation of epithelial cells in a serum-rich medium promotes EMT (Dumont et al., 2008). As pointed out above, human epidermal keratinocytes cultured *in vitro* express vimentin (Van Muijen et al., 1987; Richard et al., 1990; Johnson & Spandau, 1997) and **fibronectin (Brown & Parkinson, 1983; O'Keefe et al., 1984), typical** biomarkers of cells undergoing EMT.

As mentioned above, a link between the EMT and the gain of epithelial stem cell properties was suggested. Interestingly, primary epidermal keratinocytes plated at clonal density produced holoclones (Barrandon & Green, 1987) that had the greatest reproductive capacity and are likely to be pluripotent cells with stem-cell properties generated during partial EMT.

This data demonstrates that in culture adult epidermal keratinocytes remain responsive to the cues from mesenchymal cells and become stem cell-like cells. It is important to note that cultured keratinocyte sheets grafted to full-thickness wounds differentiate and re-express the original phenotypes (Compton et al., 1989; 1998), probably through the MET process.

We suggest that, in the presence of fibroblasts or some growth factors, cultured human epithelial cells acquire an EMT-like intermediate phenotype allowing cell migration and expansion. The great success of regenerative medicine was due to the EMT phenomenon in cultured epidermal cells and the MET process after a back transplantation of cultured epithelial sheets.

Conclusion

EMT, a long-term continuous process, is characterized by multiple morphological, metabolic, and behavioral changes in epithelial cells that enable them to assume a mesenchymal cell phenotype. Depending on the microenvironment, epithelial cells may advance through the EMT process and stall at different intermediate states. This is the way epithelial cells can achieve tremendous plasticity, which depends on the cross talk of a plethora of intercellular signaling pathways and environmental cues. The epidermal cell phenotype of an intact epidermis dramatically changes after wounding or cultivation *in vitro*. Keratinocytes undergo activation resulting in elevated proliferation, changes of keratin expression and cell-cell adhesion, and gain mesenchymal features including enhanced motility and cytoskeletal rearrangements. On a plastic substrate, epidermal cells form multicellular colonies manifesting collective migration when the coordinated dislocation of ten thousands cells takes place along the substrate. In cell culture and in wound healing, adult human epidermal keratinocytes reveal morphological and metabolic alterations that may be interpreted as EMT biomarkers. So far as it has been studied, it is possible to suggest that epidermal keratinocytes demonstrate *bona fide* partial EMT during collective migration and scratch wound repair.

EMT as a main program of plasticity of epidermal cells is of paramount importance in skin development and wound healing, and may be utilized in regenerative medicine when cultured keratinocyte sheets suitable for grafting are produced. Adult epidermal keratinocytes have not irreversibly lost the capacity to organize into epithelial tubules, but retain a latent tubulogenic potential that can be unmasked by specific signals in an *in vitro* situation. Understanding tubular morphogenesis is also important for gaining insights into the etiology of skin diseases and potentially their treatment by regenerative medicine approaches.

EMT in hair follicles during development and hair cycling is poorly understood. Down-regulation of E-cadherin and increased Snai2 expression during budding morphogenesis and in postnatal hair follicles temporally and spatially give grounds for the conclusion that elements of partial EMT may be seen during development of the epidermis and its adnexa . One may speculate, that the partial EMT process is an integral part of regularly repeating the hair growth cycle.

The mechanisms underlying the epigenetic regulation of EMT pave a way for the development of new treatment strategies for hair follicle disorders and cutaneous malignant tumors. The progress in our understanding of the EMT phenomena involved in hair follicle development and function may eventually have therapeutic implications.

References

Acloque H., Adams M. S., Fishwick K., Bronner-Fraser M., Nieto M. A. Epithelial-mesenchymal transition: the importance of changing cell state in development and disease. *J. Clin. Invest.* 2009; 119: 1438-1449.

Akhurst R. J, Derynck R: TGF-beta signaling in cancer: a doubleedged sword. *Trends Cell Biol.* 2001; 11: S44-S51.

Arnoux V., Côme C., Kusevitt D. F. et al. Cutaneous wound healing: a partial and reversible EMT. In Savagner P (ed.), Rise and Fall of Epithelial Phenotype: Concepts of Epithelial-Mesenchymal Transition. Austin, TX; Landes Biosciences. 2004; ch.8.

Arwert E. N., Hoste E., Watt F. M. Epithelial stem cells, wound healing and cancer. *Nature Reviews Cancer*. 2012; 12:170-180.

Altan Z. M., Fenteany G. c-Jun N-terminal kinase regulates lamellipodial protrusion and cell sheet migration during epithelial wound closure by a gene expression-independent mechanism. *Biochem. Biophys. Res. Commun.* 2004; 322: 56-67.

Balkovetz D. F, Pollack A. L., Mostov K. E. Hepatocyte growth factor alters the polarity of Madin–Darby canine kidney cell monolayers. *J. Biol. Chem.* 1997; 272: 3471–3477.

Barrandon Y., Green H. Three clonal types of keratinocyte with different capacities for multiplication. *Proc. Natl. Acad. Sci. USA*. 1987; 84: 2302-2306.

Batlle E., Sancho E., Tranci C., Dominguez D., Monfar M., Baulida J., Garcia De Herreros A. The transcription factor snail is a repressor of E-cadherin gene expression in epithelial tumour cells. *Nat. Cell Biol.* 2000;2: 84-89.

Beitel G. J., Krasnow M. A. Genetic control of epithelial tube size in the Drosophila tracheal system. *Development*. 2000; 127: 3271-3282.

Birchmeier C., Birchmeier W., Gherardi E., Vande Woude G. F. *Nat. Rev. Mol. Cell Biol.* 2003; 4: 915-925.

Bolós V., Peinado H., Perez-Moreno M. A., Fraga M. E., Esteller M., Cano A. The transcription factor Slug represses E-cadherin expression and induces epithelial to mesenchymal transitions: a comparison with Snail and E47 repressors. *J. Cell Sci.* 2003; 116: 499-511.

Botchkarev V. A., Kishimoto J. Molecular control of epithelial-mesenchymal interactions during hair follicle cycling. *J. Invest. Dermatol. Symposium Proc.* 2003; 8: 46-55.

Botchkarev V. A., Gdula1 M. R., Mardaryev A. N., Sharov A. A., Fessing M. Y. Epigenetic regulation of gene expression in keratinocytes. *J. Invest. Dermatol.* 2012; 32: 2505-2521.

Boyer B., Tucker G. C., Ana Maria Vallés A. M., Franke W. W., Thiery J. P. Rearrangements of desmosomal and cytoskeletal proteins during the transition from epithelial to fibroblastoid organization in cultured rat bladder carcinoma cells. *J. Cell Biol.* 1989; 109: 1495-1509.

Brinkmann V., Foroutan H., Sachs M., Weidner K. M., Birchmeier W. Hepatocyte growth factor/Scatter Factor Induces a variety of tissue-specific morphogenic programs in epithelial cells. *J. Cell Biol.* 1995; 131: 1573–1586.

Brown K. W., Parkinson E. E. Glycoproteins and glycosaminoglycans of cultured normal human epidermal keratinocytes. *J. Cell Sci.* 1983; 61: 325-338.

Burk U., Schubert J., Wellner U., Schmalhofer O., Vincan E., Spaderma S., Brabletz T. A reciprocal repression between ZEB1 and members of the miR-200 family promotes EMT and invasion in cancer cells. *EMBO Rep.* 2008; 9: 582–589.

Cano A., Pérez-Moreno M. A., Roderigo I., Locassio A., Blanco M. J., del Barrio M. G., Portillo F., Nieto M. A. The transcription factor snail controls epithelial-mesenchymal transitions by repressing E-cadherin expression. *Nat. Cell Biol.* 2000; 2: 76-83.

Chen C, Wei Y, Hummel M, Hoffmann T. K, Gross M, et al. (2011) Evidence for epithelial-mesenchymal transition in cancer stem cells of head and neck squamous cell carcinoma. *PLoS ONE.* 2011;6(1): e16466.

Chermnykh E. S., Vorotelyak E. A., Gnedeva K. Y. Moldaver M. V., Yegorov E. E., Vasiliev A. V., Terskikh V. V. Dermal papilla cells induce keratinocyte tubulogenesis in culture. Histochem. *Cell Biol.* 2010; 133: 567-576.

Chmielowiec J., Borowiak M., Morkel M., Stradal T., Munz B., Werner S., Wehland J., Birchmeier C., Birchmeier W. c-Met is essential for wound healing in the skin. *J. Cell Biol.* 2007; 177: 151-162.

Compton C C, Gill JM, Bradford D. A, Regauer S , Gallico G. G, O'Connor N. E. Skin regenerated from cultured epithelial autografts on full-thickness burn wounds from 6 days to 5 years after grafting. A light, electron microscopic and immunochemical study. *Lab. Invest.* 1989; 60: 600-612.

Compton C. C., Nadire K. B., Regauer S., Simon M., Warland G., Nicolas E. O'Connor N. E., Gallico G. G., Landry D. B. Cultured human sole-derived keratinocyte grafts re-express site-specific differentiation after transplantation. *Differentiation.* 1998; 64: 45–53.

Cotsarelis, G., Sun, T. T. and Lavker, R. M. Label-retaining cells reside in the bulge area of pilosebaceous unit: implications for follicular stem cells, hair cycle, and skin carcinogenesis. *Cell.* 1990. 61: 1329–1337.

Cunha G. R., Bigsby R. M., Cooke P. S., Sugimura Y. Stromal-epithelial interactions in adult organs. *Cell Diff.* 1985; 17: 137-148.

Dave B., Mittal V., Tan N. M. Chang J. C. Epithelial–mesenchymal transition, cancer stem cells and treatment resistance. *Breast Cancr Res.* 2012; 14: 202-206.

Davies M., Robinson M., Smith E., Huntley S., Prime S., Paterson I. Induction of an epithelial to mesenchymal transition in human immortal and malignant keratinocytes by TGF-β1 involves MAPK, Smad and AP-1 signalling pathways. *J. Cell. Biochem.* 2005; 95: 918–931.

De Wever O., Westbroek W., Verloes A., Bloemen N., Bracke M., Gespach C., Bruyneel E., Mareel M. Critical role of N-cadherin in myofibroblast invasion and migration in vitro stimulated by colon-cancer-cell-derived TGF-beta or wounding. *J. Cell Sci.* 2004; 117: 4691-4703.

Ding W., You H., Dang H., LeBlanc F., Vivian Galicia V., Shelly C. Lu S.C., Bangyan Stiles B., Rountree C.B. Epithelial-to-mesenchymal transition of murine liver tumor cells promotes invasion. *Hepatology.* 2010; 52: 945-953.

Dumont N., Wilson M. B., Crawford Y. G., Reynolds P. A., Sigaroudinia M., Tlsty T. D. Sustained induction of epithelial to mesenchymal transition activates DNA methylation of genes silenced in basal-like breast cancers. *Proc. Natl. Acad. Sci.* 2008; 105: 14867-14872.

Dvorak H-F. Tumors: Wounds that do not heal. Similarities between tumor stroma generation and wound healing. *N. Engl. J. Med.* 1986; 315:1650–1659.

Eastham, A. M., Spencer, H., Soncin, F., Ritson, S., Merry, C. L., Stern, P. L., and Ward, C. M. Epithelial-Mesenchymal Transition Events during Human Embryonic Stem Cell Differentiation. *Cancer Res.* 2007; 67, 11254–11262.

Elliott, B. E., Hung, W. L., Boag, A. H., and Tuck, A. B. (2002). The role of hepatocyte growth factor (scatter factor) in epithelial-mesenchymal transition and breast cancer. *Can. J. Physiol. Pharmacol.* 80, 91–102.

Fan J. M., Ng Y. Y., Hill P. A., Nikolic-Paterson D. J., Mu W., Atkins R. C., Lan H. Y. Transforming growth factor-beta regulates tubular epithelial-myofibroblast transdifferentiation in vitro. *Kidney Int.* 1999; 56: 1455-1467.

Farooqui R., Fenteany G. Multiple rows of cells behind an epithelial wound edge extend cryptic lamellipodia to collectively drive cell-sheet movement. *J. Cell Sci.* 2005; 118: 51-63.

Foitzik K., Paus R., Doetschman T., Dotto G.P. The TGF-β2 isoform is both a required and sufficient inducer of murine hair follicle morphogenesis. *Dev. Biol.* 1999; 212: 278-289.

Franke, W. W., C. Grund, C. Kuhn, B. W. Jackson, and K. Illmansee. Formation of cytoskeletal elements during mouse embryogenesis. III. Primary mesenchymal cells and the first appearance of vimentin filaments. *Differentiation.* 1982; 23:43-59.

Fujie T, Katoh S, Oura H, Urano Y, Arase S. The chemotactic effect of a dermal papilla cell-derived factor on outer root sheath cells. *J. Dermatol. Sci.* 2001;.25:.206–212.

Gibson W. T., Couchman J. R., Badley R. A., Saunders H. J., Smith C. G. Fibronectin in cultured rat keratinocytes: distribution, synthesis, and relationship to cytoskeletal proteins. *Eur. J. Cell Biol.* 1983 30: 205-213.

Gilles C., Polette M., Zahm J.- M., Tournier J.- M. , Volders L., Foidart J.- M., Birembaut P. Vimentin contributes to human mammary epithelial cell migration. *J. Cell Science.* 1999; 112, 4615-4625.

Gnedeva K. Yu., Chermnykh E. S., Vorotelyak E. A., Vasil'ev A. V., Terskikh V. V. Effect of growth factors on morphogenesis of human keratinocytes in vitro. *Biol. Bulletin.* 2009; 36: 307–310.

Greenburg G., Hay E. D. Epithelia suspended in collagen gels can lose characteristics of migrating mesenchymal cells polarity and express characteristics of migrating mesenchymal cells. *J. Cell Biol.* 1982; 95: 333-339.

Greenburg G., Hay E. D. Cytodifferentiation and tissue phenotype change during transformation of embryonic lens epithelium to mesenchyme-like cells in vitro. *Dev. Biol.* 1986; 115: 363-379.

Gregory P. A., Bracken C. P., Smith E., Bert A. G., Wright J. A., Roslan S., Morris M., Wyatt L., Farshid G., Lim Y. Y., Lindeman G. J., Shannon M. F., Drew P. A., Khew-Goodall Y., Goodall G. J. An autocrine TGF-beta/ZEB/miR-200 signaling network regulates establishment and maintenance of epithelial-mesenchymal transition. *Mol. Biol. Cell.* 2011; 22: 1686-1698.

Guarino M., Rubino B., Ballabio G. The role of EMT in cancer pathology. *Pathology.* 2007; 39: 305-318.

Haake A. R., Lane A. T. Retention of differentiated characteristics in human fetal keratinocytes in vitro. *In Vitro Cell Dev. Biol.* 1989; 25: 592-600.

Hay, E. D. 1968. Organization and fine structure of epithelium and mesenchyme in the developing chick embryo. In Epithelial-mesenchymal interactions. R. Fleischmajer and R.E. Billingham, editors. Williams and Wilkins. Baltimore, Maryland, USA. 31–55.

Hay E. D. An overview of epithelio-mesenchymal transformation. *Acta Anat.* (Basel). 1995; 154: 8-20.

Hirai Y., Nose A., Kobayashi S., Takeichi M. Expression and role of E- and P-cadherin adhesion molecules in embryonic histogenesis. II. Skin morphogenesis. *Development.*1989; 105: 271-277.

Hogan B. L. Morphogenesis. *Cell.* 1999; 96: 225–233.

Hudson L. G., newkirk K. M., Chandler H. L., Choi C., Fossey S. L., Parent A. E., Kusewitt D. F. Cutaneous wound reepithelialization is compromised in mice lacking functional Slug (Snai2). *J. Dermatol. Sci.* 2009 56: 19-26.

Hugo H, Ackland M. L, Blick T, Lawrence M. G, Clements J. A, Williams E. D, Thompson E. W: Epithelial–mesenchymal and mesenchymal–epithelial transitions in carcinoma progression. *J. Cell Physiol.* 2007; 213:374-383.

Hutchin, M. E., Kariapper, M. S., Grachtchouk, M., Wang A., Wei L., Cummings D., Liu J., Michael L.E., Glick A., Dlugosz A.A. Sustained Hedgehog signaling is required for basal cell carcinoma proliferation and survival: Conditional skin tumorigenesis recapitulates the hair growth cycle. *Genes and Development.* 2005; 19: 214–223.

Imamichi Y., Menke A. Signaling Pathways Involved in Collagen-Induced Disruption of the E-Cadherin Complex during Epithelial-Mesenchymal Transition *Cells Tissues Organs* 2007;185:180-190

Iwano M., Plieth D., Danoff T. M., Xue C., Okada H., Neilson E.G. Evidence that fibroblasts derive from epithelium during tissue fibrosis. *J. Clin. Invest.* 2002; 110: 341–350.

Jamora C., DasGupta R., Kocieniewski P., Fuchs E. Links between signal transduction, transcription and adhesion in epithelial bud development. *Nature.* 2003; 422(6929): 317–322.

Jamora C, Lee P, Kocieniewski P, Azhar M, Hosokawa R, Chai Y., Fuchs E. A signaling pathway involving TGF-β2 and Snail in hair follicle morphogenesis. *PLoS Biol.* 2005; 3(1): e11.

Jiang S.-T., Chuang W.-J. Tang M.-J.Role of fibronectin deposition in branching morphogenesis of Madin-Darby canine kidney cells. *Kidney International.* 2000; 57: 1860–1867.

Jindo T., Tsuboi R., Imai R., Takamori K., Rubin J. S., Ogawa H. Hepatocyte growth factor/scatter factor stimulates hair growth of mouse vibrissae in organ culture. *J. Invest. Dermatol.* 1994; 103: 306-309.

Johnson S. C., Spandau D. F. Evidence that vimentin and cytokeratins are co-expressed by epidermal keratinocytes during wound healing. *J. Invest. Dermatol.* 1997, 108: 605. (abstr).

Jordan N. V., Johnson G. L., Abell A. N. Tracking the intermediate stages of epithelial-mesenchymal transition in epithelial stem cells and cancer. *Cell Cycle.* 2011; 10:17, 2865-2873.

Kalluri R., Neilson E. G. Epithelial-mesenchymal transition and its implication for fibrosis. *J. Clin. Invest.* 2003; 112: 1776-1784.

Kalluri R., Weinberg R. A. The basics of epithelial-mesenchymal transition. *J. Clin. Invest.* 2009; 119: 1420–1428.

Korpal M, Lee E. S, Hu G and Kang Y. The miR-200 family inhibits epithelial-mesenchymal transition and cancer cell migration by direct targeting of E-cadherin transcriptional repressors ZEB1 and ZEB2. *J. Biol. Chem.* 2008; 283: 14910-14914.

Koster M. I., Kim S., Mills A. A., DeMayo F. J., Roop D. R. p63 is the molecular switch for initiation of an epithelial stratification program. *Genes Dev.* 2004; 18: 126-131.

Kubo M., Norris D. A., Howell S. E., Ryan S. R., Clark R. A. Human keratinocytes synthesize, secrete, and deposit fibronectin in the pericellular matrix. *J. Invest. Dermatol.* 1984; 82:580-586.

Laberge R.-M., Awad P., Campisi J., Desprez P.-Y. Epithelial-mesenchymal transition induced by senescent fibroblasts. *Cancer Microenvironment.* 2012;5: 39-44.

Larue L., Bellacosa A. Epithelial–mesenchymal transition in development and cancer: role of phosphatidylinositol 30 kinase/AKT pathways *Oncogene.* 2005; 24: 7443-7454.

Lee J. M., Dedhar S., Kalluri R., Thompson E. W. The epithelial-mesenchymal transition: new insights in signaling, development, and disease. *J. Cell Biol.* 2006; 72: 73–981.

Lee Y. I., Kwon Y. J., Joo C. K. Integrin-linked kinase function is required for transforming growth factor beta-mediated epithelial to mesenchymal transition. *Biochem. Biophys. Res. Commun.* 2004; 16: 997-1001.

Leroy P., Mostov K. E. Slug Is Required for Cell Survival during Partial Epithelial-Mesenchymal Transition of HGF-induced Tubulogenesis. *Mol. Biol. Cell.* 2007; 18: 1943–1952.

Li Y., Yang J., Dai C., Wu C., Liu Y. Role for integrin-linked kinase in mediating tubular epithelial to mesenchymal transition and renal interstitial fibrogenesis. *J. Clin. Invest.* 2003; 112: 503–516.

Liang C.-C., Chen H.-C. Sustained activation of extracellular signal-regulated kinase stimulated by hepatocyte growth factor leads to integrin $\alpha 2$ expression that is involved in cell scattering. *J. Biol. Chem.* 2001; 276: 21146-21152.

Lindner G., Menrad A., Gehrardi E., Merlino G., Welker P., Handjiski B., Roloff B. Paus R. Involvement of hepatocyte growth factor/scatter factor and Met receptor signaling in hair follicle morphogenesis and cycling. *FASEB J.* 2000; 14: 319-332.

Maeda M, Johnson K. R., Wheelock M. J. Cadherin switching: essential for behavioral but not morphological changes during an epithelium-to-mesenchyme transition. *J. Cell Sci.* 2005; 118: 873-887.

Mani S. A., Guo W., Liao M.-J., Eaton E. Ng., Ayyanan A., Zhou A. Y., Brooks M., Reinhard F., Zhang C. C., Michail Shipitsin M., Campbell L. L., Polyak K., Brisken C., Jing Yang J., Weinberg R. A. The epithelial-mesenchymal transition generates cells with properties of stem cells. *Cell.* 2008; 133: 704-715.

Martin P., Parkhurst S. M. Parallels between tissue repair and embryo morphogenesis. *Development.* 2004; 131:3021-3034.

McDonald O. G., Wu H., Timp W., Doi A., Feinberg A. P. Genome-scale epigenetic reprogramming during epithelial to mesenchymal transition. *Nat. Struct. Mol. Biol.* 2011; 18: 867-874.

Miettinen P. J, Ebner R, Lopez A. R, Derynck R. TGF- β -induced transdifferentiation of mammary epithelial cells to mesenchymal cells: involvement of type I receptors. *J. Cell Biol.* 1994; 127: 2021–2036.

Moll R., Mitze M., Frixen U. H., Birchmeier W. Differential loss of E-cadherin expression in infiltrating ductal and lobular breast carcinomas. *Am. J. Pathol.* 1993; 143: 1731-1742.

Montesano R., Matsumoto K., Nakamura T., Orci L. Identification of a fibroblast-derived epithelial morphogen as hepatocyte growth factor. *Cell.* 1991a; 67, 901–908.

Montesano, R., Schaller, G., and Orci, L. Induction of epithelial tubular morphogenesis in vitro **by fibroblast**-derived soluble factors. *Cell.* 1991b; 66, 697–711.

Montesano R., Soriano J. V., Malinda K. M., Ponce M. L., Bafico A., Kleinman H. K.,Bottaro D. P., Aaronson S. A. Differential effects of hepatocyte growth factor isoforms on epithelial and endothelial tubulogenesis. *Cell Growth Differentiation.* 1998; 9: 355-365.

Morel A., Lièvre M., Thomas C., Hinkal G., Ansieau S, Puisieux A. Generation of breast cancer stem cells through epithelial-mesenchymal transition. *PLoS One.* 2008; 3(8): e2888.

Morris, R. J., Tryson, K., Wu, K. Q. Evidence that epidermal targets of carcinogen action are found in the interfollicular epidermis or infundibulum as well as in the hair follicles. *Cancer Res.* 2000; 60: 226–229.

Moustakas A, Heldin CH: Signaling networks guiding epithelial mesenchymal transitions during embryogenesis and cancer progression. *Cancer Sci.* 2007, 98:1512-1520.

Nakajima, Y., Yamagishi, T., Hokari, S., Nakamura, H. Mechanisms involved in valvuloseptal endocardial cushion formation in early cardiogenesis:

roles of transforming growth factor (TGF)-beta and bone morphogenetic protein (BMP). *Anat. Rec.* 2000; 258, 119–127.

Nakamura, T., K. Nawa, A. Isbihara, N. Kaise, and T. Nishino. 1987. Purification and subunit structure of hepatocyte growth factor from rat platelcts. *FEBS Lett.* 224:311-316.

Neilson E. G. Plasticity, nuclear diapause, and a requiem for the terminal differentiation of epithelia. *J. Am. Soc. Nephrol.* 2007; 18: 1995–1998.

Nieman M. T., Prudoff R. S., Johnson K. R., Wheelock M. J. N-cadherin promotes motility in human breast cancer cells regardless of their E-cadherin expression. *J. Cell Biol.* 1999; 147, 631-644.

Niessen C. M, Gottardi C. J. Molecular components of the adherens junction. *Biochim. Biophys. Acta.* 2008; 1778: 562–571.

O'Brien L. E., Tang K., Kats E. S., Schutz-Geschwender A., Lipschutz J. H., Mostov K. E. ERK and MMPs sequentially regulate distinct stages of epithelial tubule development. *Developmental Cell.* 2004; 7: 21–32.

Oh J. E., Kim R. H., Shim K. H., Park N. H., Kang M. K. ΔNp63α protein triggers epithelial-mesenchymal transition and confers stem cell properties in normal human keratinocytes. *J. Biol. Chem.* 2011; 286: 3875-3867.

Okada H., Strutz F., Danoff T. M., Kalluri R., Neilson E. G. Possible mechanisms of renal fibrosis. *Contrib. Nephrol.* 1996; 118: 147-154.

O'Keefe E. J., Woodley D., Castillo G., Russell N., Payne R. E. Production of soluble and cell-associated fibronectin by cultured keratinocytes. *J. Invest. Dermatol.* 1984; 82: 150-155.

Parent A. E., Newkirk K. M., Kusewitt D. F. Slug (Snai2) expression during skin and hair follicle development. *J. Invest. Dermatol.* 2010; 130: 1737–1739.

Peinado H, Ballestar E, Esteller M, Cano A. Snail mediates E-cadherin repression by the recruitment of the Sin3A/histone deacetylase 1 (HDAC1)/HDAC2 complex. *Mol. Cell Biol.* 2004; 24: 306–19.

Peinado H, Quintanilla M, Cano A. Transforming growth factor beta-1 induces snail transcription factor in epithelial cell lines: Mechanisms for epithelial mesenchymal transitions. *J. Biol. Chem.* 2003. 278: 21113–21123.

Pollack A. L., Runyan R. B., Mostov K. E. Morphogenetic mechanisms of epithelial tubulogenesis: MDCK cell polarity is transiently rearranged without loss of cell–cell contact during scatter factor/hepatocyte growth factor-induced tubulogenesis. *Dev. Biol.* 1998;204: 64-79.

Poujade M., Grasland-Mongrain E., Hertzog A., Jouanneau J., Chavrier P., Ladoux B., Buguin A., Silberzan P. Collective migration of an epithelial monolayer in response to a model wound. *Proc. Natl. Acad. Sci. USA.* 2007; 104; 15988-15993.

Radisky D. C, Kenny P. A, Bissell M. J. Fibrosis and cancer: Do myofibroblasts come also from epithelial cells via EMT? *J. Cell Biochem.* 2007; 101: 830–839.

Ramirez R., Hsu D., Patel A., Fenton C., Dinauer C. R., Tuttle M., Francis G. L. Over-expression of hepatocyte growth factor/scatter factor (HGF/SF) and the HGF/SF receptor (cMET) are associated with a high risk of metastasis and recurrence for children and young adults with papillary thyroid carcinoma *Clinical Endocrinology.* 2000; 53: 635-644.

Rendl M, Lewis L, Fuchs E. Molecular dissection of mesenchymal-epithelial interactions in the hair follicle. *PloS Biol.* 2005; 3:e331.

Rendl M, Polak L, Fuchs E. BMP signaling in dermal papilla cells is required for their hair follicle-inductive properties.*Genes Dev.* 2008; 22:543–557.

Rheinwald J. G, Green H. 1975. Serial cultivation of strains of human epidermal keratinocytes: the formation of keratinizing colonies from single cells. *Cell.* 6:331–343.

Richard M. H., J. Viac J., Reano A., Gaucherand M., Thivolet J. Vimentine expression in normal human keratinocytes grown in serum-free defined MCDS 153 medium. *Archives Dermatol. Res.* 1990; 282: 512-515.

Riss J., Khanna C., Koo S., et al. Cancers as wounds that do not heal: Differences and Similarities between Renal Regeneration/Repair and Renal Cell Carcinoma.*Cancer Res.* 2006; 66: 7216-7224.

Samavarchi-Tehrani P., Golipour A., David L., Sung H., Beyer T. A., Datti A., Woltjen K., Nagy A., Wrana J. L. Functional genomics reveals a BMP-driven mesenchymal-to-epithelial transition in the initiation of somatic cell reprogramming. *Cell Stem Cell.* 2010; 7: 64-77.

Sauka-Spengler T., Bronner-Fraser M. A gene regulatory network orchestrates neural crest formation. *Nat. Rev. Mol. Cell Biol.* 2008; 9:557–568.

Savagner P. Leaving the neighborhood: molecular mechanisms involved during epithelial-mesenchymal transition. *Bioessays.* 2001; 23: 912–923.

Savagner P. The epithelial–mesenchymal transition (EMT) phenomenon. *Annals Oncol.* 2010; 21(suppl.7): vii89–vii92.

Sawyer R. H., Fallon J. F. Epithelial-mesenchymal interactions in development. 1983. New York, Praeger Publishers.

Schäfer M., Werner S. Cancer as an overhealing wound: An old hypothesis revisited. *Nat. Rev. Mol. Cell Biol.* 2008; 9: 628–638.

Scheel C., Eaton E. N., Li S.H.-J., Chaffer C. L., Reinhardt F., Kah K.-J., Bell G., Guo W., Rubin J., Richardson A. L., Weinberg R.A. Paracrine and autocrine signals induce and maintain mesenchymal and stem cell states in the breast. *Cell.* 2011; 145:926-940.

Shimaoka S., Imai R., Ogawa H. Dermal papilla cells express hepatocyte growth factor. *J. Dermatol. Sci.* 1994; 7 Suppl: S79-S83.

Shimaoka S., Tsuboi R., Jindo T., Imai R., Takamori K., Rubin J. S., Ogawa H. Hepatocyte growth factor/ scatter factor expressed in follicular papilla cells stimulates human hair growth in vitro. *J. Cell. Physiol.* 1995; 165: 333-338.

Shinin V. V., Chernaya O. G., Terskikh V.V. The role of cell proliferation in tubulogenesis of human keratinocytes. (Article in Russian). *Izv. Akad. Nauk Ser. Biol.* 2002; Jul-Aug;(4):407-20.

Sonnenberg E, Meyer D, Weidner K. M, Birchmeier C. Scatter factor/ hepatocyte growth factor and its receptor, the c-met tyrosine kinase, can mediate a signal exchange between mesenchyme and epithelia during mouse development. *J. Cell. Biol.* 1993; 123: 223–235.

Stadler S. C., Allis C. D. Linking epithelial-to-mesenchymal-transition and epigenetic modifications. *Semin. Cancer Biol.* 2012; 22: 404-410.

Stockinger A., Eger A., Wolf J., Beug H., Foisner R. E-cadherin regulates cell growth by modulating proliferation-dependent β-catenin transcriptional activity. *J. Cell Biol.* 2001; 154: 1185-1196.

Stoker M., Gherardi E., Perryman M., Gray J. Scatter factor is a fibroblast-derived modulator of epithelial cell motility. Nature 1987. 327:239-242.

Stuart K. A., Riordan S. M., Lidder S., Crostella L., Williams R., Skouteris G. G. Hepatocyte growth factor/scatter factor-induced intracellular signalling. *Int. J. Exp. Pathol.* 2000; 8: 17-30.

Talbot L. J., Bhattacharya S. D., Kuo P. C. Epithelial-mesenchymal transition, the tumor microenvironment, and metastatic behavior of epithelial malignancies *Int. J. Biochem. Mol. Biol.* 2012; 3 :117-136.

Thiery J. P, Acloque H, Huang R. Y, Nieto M. A (2009) Epithelial-mesenchymal transitions in development and disease. *Cell.* 139: 871-890.

Thiery J. P., Sleeman J. P. Complex networks orchestrate epithelial-mesenchymal transitions. *Nature Rev. Mol. Cell Biol.* 2006; 7: 131–142.

Thompson E. W, Newgreen D. F, Tarin D: Carcinoma invasion and metastasis: a role for epithelial-mesenchymal transition? *Cancer Res.* 2005; 65:5991-5995.

Todaro G, Green H. Quantitative studies of the growth of mouse embyro cells in culture and their development into established lines. *J. Cell Biol.* 1963; 17:299–313.

Trinkaus J. P. Cells into Organs: The Forces that Shape the Embryo. Englewood Cliffs, NJ; Prentice Hall. 1984; 237p.

Vanderburg C. R., Hay E. D. Embryonic corneal fibroblasts transfected with E-cadherin cDNA undergo mesenchymal-epithelial transformation. *Mol. Biol. Cell.* 1992; 3:5a.

Van Muijen G. N., Wanaar S. O., Ponec M. Differentiation-related changes of cytokeratin expression in cultured keratinocytes and in fetal, newborn, and adult epidermis.*Exp. Cell Res.* 1987; 17: 331-345.

Venkov C., Plieth D., Ni T., Karmaker A., Bian A., George A. L., Neilson E. G. Transcriptional networks in epithelial-mesenchymal transition. *PloS One.* 2011; 6(9): e25354.

Vorotelyak E. A., Satdykova G. P., Vasiliev A. V., Terskikh V.V. Electron microscope study of migrating keratinocyte colonies. *Biol. Bulletin.* 1996; 23: 403-406.

Wakefield L. M., Roberts A. B. TGF-beta signaling: positive and negative effects on tumorigenesis. *Curr. Opin. Genet. Dev.* 2002; 12: 22-29.

Weidner R. M. et al., Molecular characteristics of HGF-SF and its receptor, the c-Met tyrosinekinase. *Trends Cell Biol.* 1998; 8: 404-410.

Wheelock M., Shintani Y., Maeda M., Fukumoto Y., Johnson K. R. Cadherin switching. *J. Cell Sci.* 2008; 121: 727-735.

Williams M. J., Clark P. Microscopic analysis of the cellular events during scatter factor/hepatocyte growth factor-induced epithelial tubulogenesis. *J. Anat.,* 2003; 203:483-503.

Wong S. Y., Reiter J. F. Wounding mobilizes hair follicle stem cells to form tumors. *Proc. Natl. Acad. Sci. USA.*2011; 108: 4093-4098.

Woolf A. S., Kolatsi-Joannou M., Hardman P., Andermarcher E., Moorby C., Fine L. G., Jat P. S., Noble M. D., Gherardi E. Roles of hepatocyte growth factor/scatter factor and the met receptor in the early development of the metanephros. *J. Cell Biol.* 1995; 128: 171–184.

Xu J., Lamouille S., Derynck R.TGF-β-induced epithelial to mesenchymal transition. *Cell Research.* 2009; 19:156–172.

Yang C.-C., Cotsarelis G. Review of hair follicle dermal cells. *J. Dermatol. Sci.* 2010; 57 (1): 2.

Yang J., LiuY. Dissection of key events in tubular epithelial to myofibroblast transition and its implicstions in renal interstitial fibrosis. *Am. J. Patol.* 2001; 159: 1465-1475.

Yang J., Weinberg R. A. (2008). Epithelial-mesenchymal transition: at the crossroads of development and tumor metastasis. *Dev. Cell* 2008; 14, 818–829.

Yang S. P., Worlf A. C., Yuan H. T., Scott R. J., Risdon R. A., O'Hare M. J., Winyard P. J. D. Potential biological role of transforming growth factor-beta1 in human kidney malformations. *Am. J. Pathol.* 2000; 157: 633-1647.

Zeisberg M., Neilson E. G. Biomarkers for epithelial-mesenchymal transitions. *J. Clin. Invest.* 2009; 119: 1429–1437.

Zhang Y. W, Vande Woude G. F. HGF/SF-Met signaling in the control of branching morphogenesis and invasion. *J. Cell Biochem.* 2003; 88:408-417.

Zuk A., Matlin E. D., Hay E. D. Type I collagen gel induces Madin-Darby canine kidney cells to become fusiform in shape and lose apical-basal polarity. *J. Cell Biol.* 1989; 108: 903-919.

In: Keratinocytes

ISBN: 978-1-62618-798-6

Editor: Elia Ranzato

© 2013 Nova Science Publishers, Inc.

Chapter 5

Keratinocytes: Gatekeeper in Innate Defense

Yuping Lai[*]

Shanghai Key Laboratory of Regulatory Biology,
School of Life Sciences, East China Normal University,
Shanghai, P.R. China

Abstract

Keratinocytes, as a major cell type of skin epidermis, function as a gatekeeper to effectively prevent pathogen entry or long-term survival on the skin. Multiple pattern recognition receptors (PRRs) such as toll-like receptors (TLRs) are expressed on keratinocytes, which enables keratinocyte recognition of microbial invasion. The activation of PRRs in keratinocytes in turn triggers the release of soluble effectors, such as the antimicrobial peptides that rapidly repel microbial assault. This short communication focuses on describing the expression profile of PRRs in keratinocytes and antimicrobial intermediates produced by keratinocytes after PRR activation, and highlights protective roles of keratinocytes in innate defense.

[*] Email: yplai@bio.ecnu.edu.cn.

The skin is the primary interface between the host and environmental damage such as microbes. A myriad of microbes colonize on skin and make contact with keratinocytes in the epidermis.

However, despite the abundant colonization by microbes, skin normally isn't infected or inflamed. Therefore, keratinocytes, the predominant cell type in the epidermis of skin, not only play an important role in maintaining the physical barrier between the host and the environment, but also participate in cutaneous immune responses to protect skin from infection (Kupper and Fuhlbrigge, 2004; Robert and Kupper, 1999). This short communication will describe how Toll-like receptors are activated to release antimicrobial peptides against pathogens in keratinocytes, thus maintaining skin homeostasis.

Pattern-Recognition Receptors in Keratinocytes

Toll-like receptors (TLRs) are a member of the pattern-recognition receptor (PRR) family and recognize structurally conserved molecules derived from microbes that are named pathogen-associated molecular patterns (PAPMs). As a consequence of its location, keratinocytes are exposed to millions of microbes and are endowed with an array of Toll-like receptors. It has been shown that TLRs1-10 are expressed in keratinocytes (Baker et al., 2003; Kollisch et al., 2005; Lebre et al., 2007; Schauber et al., 2007). Among these TLRs, TLR2, TLR3, TLR5 are constitutively expressed in primary human keratinocytes and immortalized human keratinocyte cell line HaCat while TLR4 was only found in HaCat cells (Kollisch et al., 2005).

However, some controversial results show no surface expression of TLR2, TLR4, TLR9 in cultured keratinocytes (Curry et al., 2003). One explanation is that the expression and function of TLRs are age-dependent.

The work done by the Elbe-Burge group has shown that significantly higher mRNA expression levels of TLRs 1-5 and the TLR4 co-factor MD2 were observed in embryonic and fetal skin (specific in keratinocytes) when compared with adult skin. TLR6 was essentially equally expressed in embryonic and fetal skin while TLR7-9 were absent in adult skin, but weekly detectable in prenatal skin(Iram et al., 2012). In addition to the expression of TLRs, the function of TLRs has age-dependent changes. The best evidence is exemplified by the fact that TLR3 exhibits marked differences in the magnitude of expression and function in keratincoytes before and after birth when compared with adults. Compared to basal keratinocytes from adults, neonatal keratinocytes secrete high levels of CXCL8, CXCL10 and TNFα upon stimulation by TLR3 ligand poly(I:C) even though keratinocytes from both age groups express comparable levels of TLR3. Moreover, fetal skin keratinocytes were already able to respond to poly(I:C) with a similar strength to neonatal keratinocytes(Iram et al., 2012). The data suggests the existence of age-specific responses, rather than a global, linear progression from a prenatal to an adult pattern. However, the mechanisms underlying these differences remain unknown.

Besides TLRs, keratinocytes express other PRRs including nucleotide-binding oligomerization domain (NOD)-like receptors (NLRs), retinoic acid-inducible gene I (RIG-I)-like receptors (RLRs) and C-type lectin receptors (CLRs). NOD2 expression was increased by peptidoglycan (PGN) and one of CLRs dectin-1 was enhanced by its ligand beta-glucan in keratinocytes (Kobayashi et al., 2009). Moreover, our unpublished data show that poly(I:C) increases both TLR3 and RIG-I expression. Toll-like receptors recognize pathogens at the cell surface (TLRs 1,2,4-6, 10) or within the endosome (TLRs 3, 7-9), whereas NLRs and RLRs act as intracellular surveillance molecules(Kumar et al., 2009). NLRs not only function in pathogen recognition but also play a role in tissue homeostasis(Kufer and Sansonetti, 2011). RLRs are crucial for host antiviral defense and sense double-stranded (ds) RNA. Viral dsRNA or synthetic ds RNA poly(I:C) is recognized by TLR3 and MDA5. The activation of these PRRs by PAMPs or endogenous signals of injury enables keratinocytes rapidly produce cytokines, antimicrobial peptides or antimicrobial intermediates (e.g. radical oxygen species and nitric oxide) in response to those damages.

Antimicrobial Intermediates in Keratinocytes

Cytokines, chemokines and antimicrobial peptides/proteins (AMPs) are major antimicrobial intermediates secreted by keratinocytes upon exposure to microbes. Upon stimulation by microbial pathogens, activation of TLRs leads to the production of interferons and inflammatory cytokines. These cytokines then activate surrounding cells to produce chemokines to recruit various inflammatory cells, such as neutrophils and macrophages, into the infected sites, thus clearing invading pathogens.

Except for cytokines and chemokines, the endogenous antimicrobial peptides secreted by keratinocytes are effector molecules of the innate host defense system (Ganz, 1999; Zasloff, 2002). Upon infection and injury, keratinocytes are able to produce AMPs within a few minutes. Around 10 antimicrobial peptides and proteins have been identified in keratinocytes, including cathelicidins (Braff et al., 2005), beta-defensins (Liu et al., 2002), regenerating islet-derived proteins (Lai et al., 2012) and others. These AMPs have a broad antimicrobial spectrum and inactivate microorganisms by direct interaction with biomembranes or other organelles. Besides their direct antimicrobial function, it has been suggested that AMPs play multiple roles as mediators of inflammation with impact on epithelial and inflammatory cells, influencing diverse processes such as cytokine release, cell proliferation, angiogenesis, wound healing, chemotaxis, immune induction, and protease antiprotease balance(Brown and Hancock, 2006; Lai et al., 2012; Yang et al., 2004).

PRRs Regulate Antimicrobial Intermediates in Keratinocytes

Human skin is exposed to millions of microbial organisms, and these microorganisms produce various kinds of PRR ligands. Keratinocytes are microbe-infected target cells and are equipped with a broad antimicrobial defense program enabling them to efficiently protect the host from microbial infection.

This protective function is partly mediated by the presence of AMPs as well as inflammatory cytokines and chemokines after PRRs have been activated in keratinocytes.

For example, peptidoglycan and lipoteichoic acid from *S. aureus* activated TLR2 and NOD2 on keratinocytes, resulting in activation of NF-κB and subsequent production of the neutrophil chemotactic factor IL-8 and iNOS (Mempel et al., 2003; Muller-Anstett et al., 2010), thus inhibiting *S.aureus* from invading. Other studies have shown that upon TLR3 ligation, fetal keratinocytes were able to produce CXCL9-11 and CCL3-5 that play a potential role in host defense against HSV-1(Iram et al., 2012; Nakayama et al., 2006). In addition to CXCL9-11 and CCL3-5, activation of TLR3 in cultured human keratinocytes induced production of TNFα, IL-18, and type I interferon (IFNα/β) and the development of Th-1 type immune responses

(Lebre et al., 2003). Moreover, our data showed that RNA from necrotic cells triggered TLR3 on undamaged keratinocytes around wounds to produce pro-inflammatory cytokines IL-6 and TNFα after a skin injury (Lai et al., 2009), thus helping to keep wounds sterile. Altogether, these data suggest that keratinocytes, via TLR activation, play an important protective role in microbial infections and skin injury.

In addition to inflammatory cytokines and chemokines, our recent data shows that expression of murine beta-defensin was upregulated by bacterial lipopeptides in wild-type keratinocytes, while it was attenuated in TLR2-deficient keratinocytes in vitro (Lai et al., 2010). In line with decreased the expression of beta-defensins, the infection of *S.aureus* in TLR2-deficient mice was increased (Lai's unpublished data). Moreover, recent work showed hormonally active vitamin D(3)-1,25- dihydroxyvitamin D(3) (1,25D3) acted as a signaling molecule in cutaneous immunity by increasing pattern recognition through Toll-like receptor-2 (TLR2), increasing the expression and function of the antimicrobial peptide cathelicidin, and killing off intracellular Mycobacterium tuberculosis (Liu et al., 2006). In addition, the Gallo group found that the epigenetic control of gene transcription by histone acetylation is important for 1,25D3-regulated antimicrobial and TLR function of keratinocytes against *S.aureus* (Schauber et al., 2008). Beside infection, skin injury also enhanced TLR2 function to induced antimicrobial peptide expression (Schauber et al., 2007), thus promoting wound healing (Lai et al., 2012).

Conclusion

Keratinocytes, the major constituents of the epidermis, produce cytokines and antimicrobial peptides/proteins via activation of PRRs right after skin is exposed to environmental damages. Keratinocytes provide a quick innate immune response to protect the host, functioning as a gatekeeper. Furthermore, the activation of PRRs in keratinocytes plays a vital role in skin infection and injury, thereby making them potential therapeutic targets. Therefore, the ability of PRRs to combat those diseases could be used in a dermatological clinic through the development of drugs that act as PRRs agonists or antagonists.

Acknowledgments

This work was supported by the National Natural Science Foundation of China grants (NSFC) 81072422, 31222021and31170867, grants NCET-11-0141, 11QA1401900, 12ZZ039 to Yuping Lai and the Science and Technology Commission of Shanghai Municipality grant 11DZ2260300. I declare no conflict of interest.

References

Baker, B.S., Ovigne, J.M., Powles, A.V., Corcoran, S., and Fry, L. (2003). Normal keratinocytes express Toll-like receptors (TLRs) 1, 2 and 5: modulation of TLR expression in chronic plaque psoriasis. *Br. J. Dermatol. 148*, 670-679.

Braff, M.H., Zaiou, M., Fierer, J., Nizet, V., and Gallo, R.L. (2005). Keratinocyte production of cathelicidin provides direct activity against bacterial skin pathogens. *Infect. Immun. 73*, 6771-6781.

Brown, K.L., and Hancock, R.E. (2006). Cationic host defense (antimicrobial) peptides. *Curr. Opin. Immunol. 18*, 24-30.

Curry, J.L., Qin, J.Z., Bonish, B., Carrick, R., Bacon, P., Panella, J., Robinson, J., and Nickoloff, B.J. (2003). Innate immune-related receptors in normal and psoriatic skin. *Arch. Pathol. Lab. Med. 127*, 178-186.

Ganz, T. (1999). Defensins and host defense. Science *286*, 420-421.

Iram, N., Mildner, M., Prior, M., Petzelbauer, P., Fiala, C., Hacker, S., Schoppl, A., Tschachler, E., and Elbe-Burger, A. (2012). Age-related changes in expression and function of Toll-like receptors in human skin. *Development 139*, 4210-4219.

Kobayashi, M., Yoshiki, R., Sakabe, J., Kabashima, K., Nakamura, M., and Tokura, Y. (2009). Expression of toll-like receptor 2, NOD2 and dectin-1 and stimulatory effects of their ligands and histamine in normal human keratinocytes. *Br. J. Dermatol. 160*, 297-304.

Kollisch, G., Kalali, B.N., Voelcker, V., Wallich, R., Behrendt, H., Ring, J., Bauer, S., Jakob, T., Mempel, M., and Ollert, M. (2005). Various members of the Toll-like receptor family contribute to the innate immune response of human epidermal keratinocytes. *Immunology 114*, 531-541.

Kufer, T.A., and Sansonetti, P.J. (2011). NLR functions beyond pathogen recognition. *Nat. Immunol. 12*, 121-128.

Kumar, H., Kawai, T., and Akira, S. (2009). Pathogen recognition in the innate immune response. *Biochem. J. 420*, 1-16.

Kupper, T.S., and Fuhlbrigge, R.C. (2004). Immune surveillance in the skin: mechanisms and clinical consequences. *Nat. Rev. Immunol. 4*, 211-222.

Lai, Y., Cogen, A.L., Radek, K.A., Park, H.J., Macleod, D.T., Leichtle, A., Ryan, A.F., Di Nardo, A., and Gallo, R.L. (2010). Activation of TLR2 by a small molecule produced by Staphylococcus epidermidis increases antimicrobial defense against bacterial skin infections. *J. Invest. Dermatol. 130*, 2211-2221.

Lai, Y., Di Nardo, A., Nakatsuji, T., Leichtle, A., Yang, Y., Cogen, A.L., Wu, Z.R., Hooper, L.V., Schmidt, R.R., von Aulock, S., *et al.* (2009). Commensal bacteria regulate Toll-like receptor 3-dependent inflammation after skin injury. *Nat. Med. 15*, 1377-1382.

Lai, Y., Li, D., Li, C., Muehleisen, B., Radek, K.A., Park, H.J., Jiang, Z., Li, Z., Lei, H., Quan, Y., *et al.* (2012). The antimicrobial protein REG3A regulates keratinocyte proliferation and differentiation after skin injury. *Immunity 37*, 74-84.

Lebre, M.C., Antons, J.C., Kalinski, P., Schuitemaker, J.H., van Capel, T.M., Kapsenberg, M.L., and De Jong, E.C. (2003). Double-stranded RNA-exposed human keratinocytes promote Th1 responses by inducing a Type-1 polarized phenotype in dendritic cells: role of keratinocyte-derived tumor necrosis factor alpha, type I interferons, and interleukin-18. *J. Invest. Dermatol. 120*, 990-997.

Lebre, M.C., van der Aar, A.M., van Baarsen, L., van Capel, T.M., Schuitemaker, J.H., Kapsenberg, M.L., and de Jong, E.C. (2007). Human keratinocytes express functional Toll-like receptor 3, 4, 5, and 9. *J. Invest. Dermatol. 127*, 331-341.

Liu, A.Y., Destoumieux, D., Wong, A.V., Park, C.H., Valore, E.V., Liu, L., and Ganz, T. (2002). Human beta-defensin-2 production in keratinocytes is regulated by interleukin-1, bacteria, and the state of differentiation. *J. Invest. Dermatol. 118*, 275-281.

Liu, P.T., Stenger, S., Li, H., Wenzel, L., Tan, B.H., Krutzik, S.R., Ochoa, M.T., Schauber, J., Wu, K., Meinken, C., *et al.* (2006). Toll-like receptor triggering of a vitamin D-mediated human antimicrobial response. *Science 311*, 1770-1773.

Mempel, M., Voelcker, V., Kollisch, G., Plank, C., Rad, R., Gerhard, M., Schnopp, C., Fraunberger, P., Walli, A.K., Ring, J., *et al.* (2003). Toll-like receptor expression in human keratinocytes: nuclear factor kappaB controlled gene activation by Staphylococcus aureus is toll-like receptor 2 but not toll-like receptor 4 or platelet activating factor receptor dependent. *J. Invest. Dermatol. 121*, 1389-1396.

Muller-Anstett, M.A., Muller, P., Albrecht, T., Nega, M., Wagener, J., Gao, Q., Kaesler, S., Schaller, M., Biedermann, T., and Gotz, F. (2010). Staphylococcal peptidoglycan co-localizes with Nod2 and TLR2 and activates innate immune response via both receptors in primary murine keratinocytes. *PLoS One 5*, e13153.

Nakayama, T., Shirane, J., Hieshima, K., Shibano, M., Watanabe, M., Jin, Z., Nagakubo, D., Saito, T., Shimomura, Y., and Yoshie, O. (2006). Novel antiviral activity of chemokines. *Virology 350*, 484-492.

Robert, C., and Kupper, T.S. (1999). Inflammatory skin diseases, T cells, and immune surveillance. *N. Engl. J. Med. 341*, 1817-1828.

Schauber, J., Dorschner, R.A., Coda, A.B., Buchau, A.S., Liu, P.T., Kiken, D., Helfrich, Y.R., Kang, S., Elalieh, H.Z., Steinmeyer, A., *et al.* (2007). Injury enhances TLR2 function and antimicrobial peptide expression through a vitamin D-dependent mechanism. *J. Clin. Invest. 117*, 803-811.

Schauber, J., Oda, Y., Buchau, A.S., Yun, Q.C., Steinmeyer, A., Zugel, U., Bikle, D.D., and Gallo, R.L. (2008). Histone acetylation in keratinocytes enables control of the expression of cathelicidin and CD14 by 1,25-dihydroxyvitamin D3. *J. Invest. Dermatol. 128*, 816-824.

Yang, D., Biragyn, A., Hoover, D.M., Lubkowski, J., and Oppenheim, J.J. (2004). Multiple roles of antimicrobial defensins, cathelicidins, and eosinophil-derived neurotoxin in host defense. *Annu. Rev. Immunol. 22*, 181-215.

Zasloff, M. (2002). Antimicrobial peptides of multicellular organisms. *Nature 415*, 389-395.

In: Keratinocytes
Editor: Elia Ranzato

ISBN: 978-1-62618-798-6
© 2013 Nova Science Publishers, Inc.

Chapter 6

Mechanisms for Keratinocyte Immune Defense: Implications for Cutaneous Wound Healing

Juan L. Rendon and Katherine A. Radek[*]
Loyola University Chicago, Health Sciences Campus, Department of
Surgery, Maywood, IL, US

Abstract

Keratinocytes comprise 95% of the epidermis and provide a barrier
to the external environment by limiting the loss of heat, moisture, and
vital constituents, while blocking the penetration of pathogens and
exogenous toxic molecules. Keratinocytes accomplish this by assembling
a unique composition of chemical, ionic, lipid, and physical barriers that
facilitate the skin's innate immune function and tissue repair pathways.
Keratinocytes function as immunomodulators that are responsible for
producing and secreting inhibitory cytokines in the absence of injury to
maintain normal epidermal barrier integrity. In addition, these cells can
promote inflammation via production of pro-inflammatory mediators and
activation of local antigen-presenting cells following tissue injury.
Activation of these pathways is regulated by various innate immune and
neuroendocrine-mediated mechanisms. Interestingly, recent

---

[*] Katherine A. Radek, Loyola University Chicago, Health Sciences Campus, Department of
Surgery, 2160 S. First Avenue, Maywood, IL 60153, USA. Email: kradek1@lumc.edu.

investigations have uncovered that commensal microflora present on healthy human skin limits the overgrowth of resident opportunistic pathogens, while simultaneously restricting the colonization of impending pathogenic microbes. This chapter highlights the key pathways that modulate keratinocyte immune defense, and the dynamics governing host-microbe interactions critical for normal cutaneous wound healing.

Keratinocyte Toll-Like Receptors Modulate Tight Junctions and Skin Immunity

In addition to providing a physical barrier, keratinocytes contribute to the immunologic barrier that has evolved to prevent invasion of pathogens. Like other cells of the innate immune system, keratinocytes utilize various pathogen recognition receptors (PRRs), including toll-like receptors (TLRs), to distinguish pathogens from host molecules via recognition of various conserved pathogen-associated molecular patterns (PAMPs). Though keratinocytes have been shown to express functional TLRs 1, 2, 3, 4, 5 and 9 [1-6], the most well described in modulation of tight junction formation and skin immunity include TLRs 2, 3 and 4 [7, 8].

Activation of TLR2, particularly by commensal bacterium *Staphylococcus epidermidis*, is of central importance in understanding the interaction between the physical and immune functions of keratinocytes. TLR2 is expressed beneath the stratum corneum, the superficial most layer of the epidermis. Upon breach of the skin barrier, TLR2 can recognize peptidoglycan present within the membrane of Gram-positive bacteria [3-5]. Experimentally, the quality of keratinocyte barrier function is determined by measuring transepithelial electric resistance (TER), where a higher TER means lower permeability of ions and greater barrier integrity. Moreover, a higher TER is associated with an enhancement of tight junction formation [9]. Mechanistically, *in vitro* studies led to the discovery that activation of TLR2 by peptidoglycan directly enhances TER by increasing associations between occludin proteins and phospho-atypical PKC iota and zeta [7]; thereby, improving the quality of the physical barrier. Atypical PKC iota and zeta strengthen tight junction barrier function by phosphorylating tight junction proteins, including occluding, Figure 1 [10, 11]. Thus, the presence of Gram-positive bacteria, including commensal *Staphylococcus epidermidis*, is critical to epidermal barrier and immune homeostasis, such that breech of the physical barrier results in activation of TLR2 and subsequent strengthening of tight junctions to prevent invasion of pathogens.

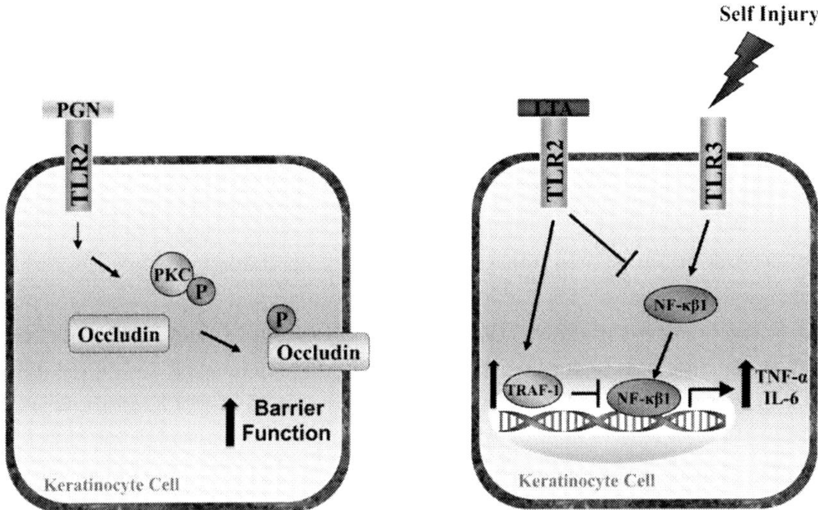

Figure 1. TLR2 regulates keratinocyte immune and barrier function. Activation of TLR2 by peptidoglycan (PGN) directly enhances barrier function by increasing associations between occludin proteins and phospho-atypical PKC, left panel. Keratinocytes also interact with commensal bacteria to maintain immune homeostasis. In the skin, injury-induced necrosis and apoptosis activate TLR3 signaling to promote expression of pro-inflammatory cytokines. However, this is regulated by TLR2, where activation by lipoteichoic acid (LTA) inhibits TLR3-dependent translocation of NF-κB1 and induces negative regulators of TLR3 signaling, including TNF receptor-associated factor-1 (TRAF-1), right panel. Under normal conditions, the skin has developed unique mechanisms to recognize injury and improve barrier function and immunity, but prevent uncontrolled inflammation.

In addition to modulating the physical barrier, keratinocytes also interact with commensal bacteria to maintain immune homeostasis at the interface between the internal and external environment. Classically, stimulation of TLRs results in activation of NF-κB and subsequent production of several pro- and anti-inflammatory cytokines. Although keratinocyte stimulation with peptidoglycan induces the expression of pro-inflammatory TNF-α and IL-6, this effect is transient and last for only a few hours [7]. In addition to peptidogylcan, the cell wall of Gram-positive bacteria (e.g. *Staphylococcus epidermidis)* contains lipoteichoic acid that functions as an alternative ligand and activator of TLR2. Furthermore, recognition of lipoteichoic acid by TLR2 on immune cells under sterile conditions promotes the induction of pro-inflammatory cytokines [12]. However, activation of keratinocytes with lipoteichoic acid obtained from *Staphylococcus epidermidis* inhibits TLR3-dependent TNF-α production [8]. Thus, the inflammatory and immune responses to TLR2 activation are ligand and environment dependent, as cutaneous stimulation with peptidoglycan promotes the inflammatory process while lipoteichoic acid inhibits inflammation. Together, these observations demonstrate that, under normal conditions, the skin immune system minimizes unnecessary inflammation but allows for an immediate response to injury and infection.

The balance between immune activation and suppression becomes markedly important in the context of skin injury, as damage to the skin barrier increases the activation of necrotic and apoptotic pathways [13, 14]. Skin injury models have correlated non-infected skins wounds with increased TNF-α and IL-6, and found this to be a TLR3-dependent process [8]. Originally, TLR3 was notable for recognition of double-stranded RNA from viruses; however, newer experimental models show that keratinocytes use TLR3 to recognize self-injury [8].

Specifically, mice lacking TLR3 do not exhibit increased levels of TNF-α or IL-6 following punch skin wounding. Similarly, *in vitro* conditions demonstrate that necrotic cells activate the expression of TNF-α from keratinocytes, a process that does not occur in keratinocytes where TLR3 is silenced [8].

The observation that skin injury activates TLR3 through recognition of self-injury, without inducing uncontrolled inflammation, can be explained by the interaction between commensal bacteria and keratinocytes. As mentioned previously, lipoteichoic acid containing *Staphylococcus epidermidis* inhibits TLR3-dependent signaling. Specifically, *Staphylococcus epidermidis* inhibits the translocation of NF-κB1 to block the expression of pro-inflammatory cytokines.

Concurrently, lipoteichoic acid induces negative regulators of TLR3 signaling, including TNF receptor-associated factor-1 (TRAF-1), Figure 1 [8]. In mice lacking TLR2 or its downstream adaptor molecule MyD88, lipoteichoic acid does not suppress TLR3-dependent cytokine activation. Thus, the skin has developed a unique mechanism by which to utilize TLR signaling to recognize injury and improve barrier function and immunity, while limiting uncontrolled inflammation.

Antimicrobial Proteins, Regulators of Skin Immunity and Barrier Function

Antimicrobial peptides (AMPs) are components of the innate immune system that are involved in preventing infection and modulating the epithelial barrier. Among the various classes of cationic AMPs produced by keratinocytes, peptidoglycan-binding proteins and S100 proteins [15]. While many AMPs are constitutively expressed, most are controlled by the local cytokine milieu. Notably, T helper cytokines IL-17 and IL-22 are key regulators of AMP expression and regulation [15]. Both IL-17 and IL-22 have been shown to increased expression of psoriasin (*S100A7*), calgranulin A (*S100A8*), calgranulin B (*S100A9*) (calgranulin A and B can form a heterdimer), human β-defensin-2 (hBD-2, *DEFB4*) and human β-defensin-3 (hBD-3, *DEFB3*) [15-19].

The peptidoglycan-binding family of AMPs includes regenerating islet-derived protein 3 (REG3). Various isoforms on REG3 have been extensively studied in the gut, where they are known to inhibit bacterial growth and support regeneration and proliferation of epithelial cells [20-23]. Not surprisingly, human REG3A (Reg3γ in mice) is highly expressed in psoriatic skin lesions as well as following injury, a process that correlates with increased levels of IL-17 and is dependent on activation of the IL-17RA receptor [24]. Unlike in the gut, where IL-22 independently induces expression of Reg3 proteins [20, 21], treatment of keratinocytes with IL-22 alone does not induce expression of REG3A. Rather, IL-22 acts in combination with IL-17 to augment the induction of REG3A [24]. Functionally, increased levels of REG3A and Reg3γ promote re-epithelization following skin injury, and maintain a proliferative response in psoriatic lesions. Though increased proliferation is necessary during wound repair, hyperproliferation is a pathologic characteristic of psoriasis. Therefore, understanding how REG3 proteins modulate the balance between proliferation and differentiation is of critical importance to many clinical conditions.

During wound healing, increased keratinocyte proliferation correlates with decreased keratinocyte differentiation [25, 26]. In line with these observations, treatment of undifferentiated keratinocytes with REG3 induces proliferation, as marked by an increased number of cells in the S phase of cell division and an inhibition of various terminal differentiation genes, including keratin-10, filaggrin and loricrin [24]. Mechanistically, REG3A and other REG proteins bind to exostosin-like 3 (EXTL3) [27, 28]. In the skin, EXTL3 is required for

REGA-induced cell proliferation via activation of phosphatidylinositol 3 kinase (PI3K) and phosphorylation of Akt [24]. Phosphorylation of Akt stimulates cellular proliferation while inhibiting differentiation to act as as a critical regulator of wound re-epithelialization.

Neuroendocrine Effects on Keratinocyte Innate Immunity and Wound Healing

The neuroendocrine axis plays a crucial role in the regulation of innate immunity and wound healing, largely through modulation of AMP activity. AMPs control immune responses, as well as facilitate several aspects of wound healing. A fundamental understanding of the interrelationship between neuroendocrine activity and AMP regulation of immunity and wound healing has emerged from observations during physiological and psychological stress. In an acute setting, wounding, pathogen challenge and stress all activate the **evolutionarily conserved autonomic "fight or flight" response**. However, prolonged activation of the stress response directly induces immunosuppression, facilitating infection and impairing wound repair processes [29-32].

Specifically, the nervous, endocrine and immune systems interact through three major pathways. The first involves activation of the autonomic system, which causes the release of catecholamines (including adrenaline and **noradrenaline, which bind α- and β-adrenergic receptors**) via aderengic stimulation. The second is through activation of the hypothalamic-pituitary (HPA) axis, which results in release of glucocorticoids (cortisol). Lastly, the third is through activation of cholinergic pathway, which culminates in the release of acetylcholine, which can activate muscarinic and nicotinic receptors [32].

In the context of psychological stress, several AMPs that normally prevent infection and improve barrier function are compromised. Cortisol secretion following activation of the HPA axis causes downregulation of cathelicidin as **well as β-defensin 3**, which subsequently impairs barrier function and facilitates cutaneous infection [29]. Similarly, activation of nicotinic receptors also impairs AMP expression and bactericidal activity. Acetylcholine, released from keratinocytes following cholinergic activation, suppresses keratinocyte antimicrobial peptide activity and barrier function [31]. In addition to directly modulating AMP activity, glucocorticoids also influence the release of acetylcholine as well as expression of nicotinic receptors [33, 34]. This phenomenon may likely further exacerbate AMP suppression, although this mechanism requires further scrutiny. Therefore, in the presence of chronic stress, heightened activation of the HPA axis and the cholinergic pathway inhibit AMP expression and activity. Ultimately, prolonged stress facilitates bacterial growth and invasion, and interferes with AMP-dependent tissue repair responses.

Psoriasis and Rosacea: Implications for Wound Healing

Pathologic skin conditions, including rosacea and psoriasis are characterized by defects in skin immunity and subsequent impaired barrier function. Rosacea is a chronic inflammatory skin disease characterized by facial flushing, erythema, inflammatory nodules, papules, pustules and telangiectasia [35]. Pathologically, rosacea is due to an inappropriate inflammatory response to normal environmental stimuli, including UV light, heat and various microbes [35]. Multiple pathways have been suggested to play a role in the development of rosacea. As mentioned above, TLR2 plays a central role in maintaining skin immune and barrier homeostasis. In patients with rosacea, there is increased expression of TLR2 on keratinocytes, a response which is augmented by elevated levels of glucocorticoids that likely occur during periods of physiological and psychological stress [36]. Moreover, TLR2 expression extends through the dermis in patients with rosacea, where it is limited to the epidermis in normal skin, Figure 2A [37]. Upregulation of TLR2 leads to uncontrolled sensitivity to environmental stimuli, including commensal bacterium *Propionibacterium*. Specifically, activation of TLR2 signaling induces post-transcriptional transport, release from keratinocyte cytoplasmic granules, and activity of the serine protease stratum corneum tryptic enzyme (SCTE; also known as kallikrein 5, KLK5 or hK5) [37, 38]. KLK5 cleaves proprotein 18-kDa cationic antimicrobial protein (CAP18) into active cathelicidin. Aberrant KLK5 activity augments the proteolytic processing of cathelicidin into smaller immunostimulatory peptides not present in normal skin. Therefore, rosacea is a disease of both increased production of abnormal cathelicidin and greater proteolytic activity. Once released from keratinocytes, LL-37 also stimulates the production of the pro-inflammatory cytokine IL-8, which promotes neutrophil infiltration [38]. Together, increased inflammatory cytokine expression and inflammatory cell infiltrate lead to several of the clinical manifestations associated with rosacea.

Table 1. Antimicrobial Peptide/Proteins role in wound healing

Antimicrobial Peptide/Protein	Alternative Names	Role in wound healing
Cathelicidin	*LL-37*	<ul><li>Regulates chemotaxis of mast cells, monocytes, T lymphocytes, and neutrophils</li><li>Promotes angiogenesis<ul><li>Transactivates epidermal growth factor (EGFR) and induces vascular endothelial growth factor (VEGF)</li><li>Induces proliferation and migration of endothelial cells</li></ul></li><li>Facilitates re-epithelialization</li><li>Induces expression of matrix components<ul><li>Heparan sulfate proteoglycans</li></ul></li></ul>
β-Defensin 2, 3, 4	hBD-2, 3, 4	<ul><li>Promote chemotaxis of dendritic cells and T cells</li><li>Promote re-epithelialization<ul><li>Induce keratinocyte proliferation and migration</li><li>Elicit intracellular calcium mobilization in keratinocytes</li><li>Active EGFR to induce phosphorylation of both STAT1 and STAT3</li></ul></li><li>Facilitate migration and proliferation of endothelial cells, independent of VEGF</li></ul>
Catestatin	Cst	<ul><li>Constitutively expressed, induced by injury</li><li>Enhances migration and proliferation of normal keratinocytes</li><li>Induced expression of IL-8</li></ul>

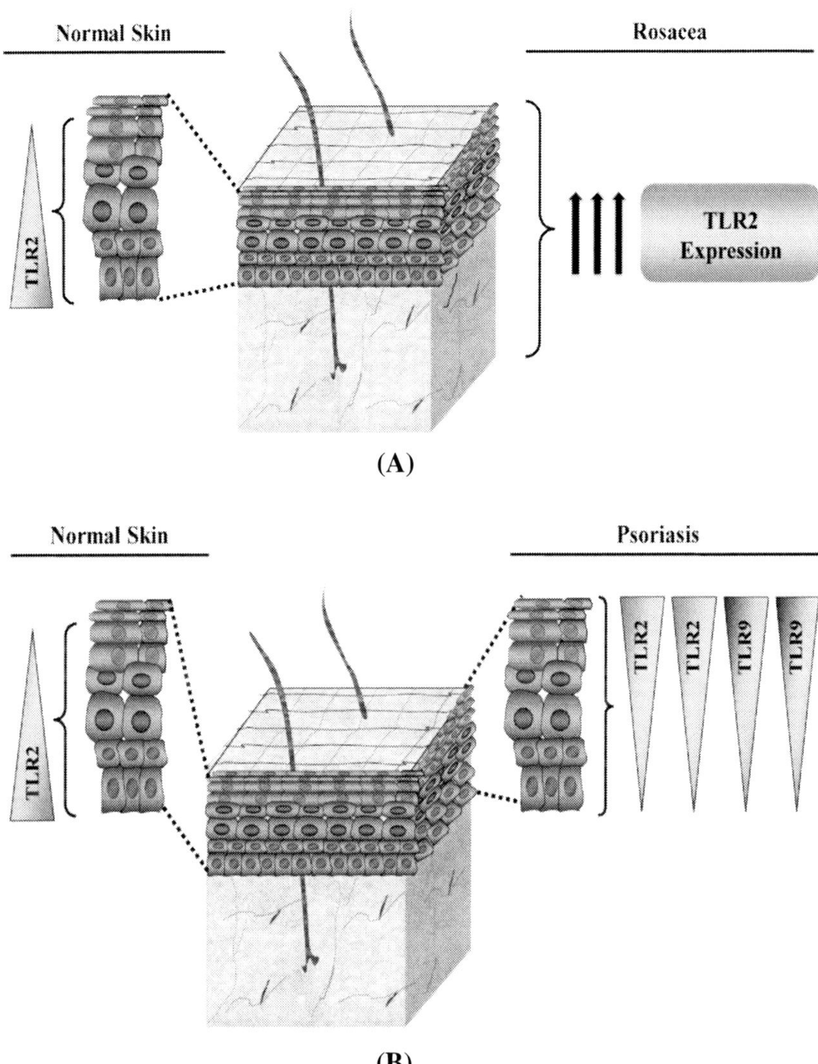

Figure 2. TLR expression in rosacea and psoriasis. In patients with rosacea there is increased expression of TLR2 on keratinocytes, which extends through the dermis, in normal skin TLR2 is limited to the epidermis, panel A. Upregulation of TLR2 leads to uncontrolled sensitivity to environmental stimuli. In psoriasis, TLR2 expression is also increased; however, while TLR2 is mostly found in the basal keratinocytes of normal skin, it is highly expressed in the upper epidermal layers of psoriatic lesions, panel B. Cathelicidin LL-37 enhances immunity in psoriatic lesions by inducing expression of TLR9 on keratinocytes.

Uncontrolled inflammation and cellular proliferation results in the characteristic hyperproliferation and lesion formation observed in psoriatic skin [18, 39]. Interestingly, patients with psoriasis also have decreased susceptibility to infection [40], yet psoriatic skin lesions exhibt higher total numbers of bacteria [41, 42]. Multiple associations have been made between the presence of specific bacteria on the skin and other areas in patients with psoriasis [43]. However, *Staphylococcus aureus* is one of the few bacteria increased in the psoriatic lesions as compared to normal skin [41]. Even so, only 20-50% of patients with psoriasis are colonized with *Staphylococcus aureus* [41, 44]. Increased total bacteria are hypothesized to play a role inducing changes in TLR and AMP expression in patients with psoriasis. As with rosacea, TLR2 expression in increased in psoriatic lesions [45]. However, while TLR2 is mostly found in the basal keratinocytes of normal skin, it is highly expressed in the upper epidermal layers of psoriatic lesions, Figure 2B [2].

Similar to rosacea, psoriasis is linked with abnormal AMP expression. In rosacea, the bioactive form of cathelicidin, LL-37, is elevated, in addition to other more pro-inflammatory cleavage products of hCAP18. The production of more immunostimulatory cleavage products of cathelicidin is due to a more diffuse and abundant expression of kallekrein proteases in the epidermis, which are known to proteolytically process and activate cathelicidin [38]. Conversely, LL-37 is the only form of cathelicidin found in psoriatic lesions [46]. Notably, LL-37 is of paramount importance to the observed augmentation of the immune response in psoriatic lesions. LL-37 induces the expression of TLR9 on keratinocytes [46] to allow keratinocytes to effectively recognize CpG sequences abundant on bacterial and viral DNA by producing type 1 IFNs to reduce the susceptibility of the skin to infection. In addition to increased cathelicidin, psoriatic lesions exhibit increased expression of hBD-2 and hBD-3, both of which were first described in lesions from psoriatic skin. Other keratinocyte derived AMPs that are increased in psoriasis include S100A7 and REG3A [24, 47]. Thus, disruption of the interrelationship among commensal bacteria, TLRs and AMP expression promotes several aspects of skin pathology. In the context of wound healing, the same mechanisms that promote disease in rosacea and psoriasis are necessary for proper wound healing. Tissue repair requires a series of events that include re-*epithelialization*, re-vascularization and remodeling in a pathogen-free environment, which heavily relies on key AMP-dependent wound healing responses.

Clinical Implications in Wound Healing

Today, recognition of the interdependence between keratinocytes and immune cells in regulating skin barrier and immune function is used to modulate various aspects of wound healing. Suppression of keratinocyte proliferation, differentiation and immune function can delay or reduce the efficiency of wound healing. Clinically, risk factors for impaired wound healing include advanced age, poor nutritional status, vascular insufficiency, infection, tobacco use, hypoxia, diabetes and immunosuppression [48-52]. In patients with these risk factors, delayed or non-healing wounds often require complex medical management, including correction of perfusion and oxygenation deficits, debridement of non-viable tissue, and treatment and prevention of infection and surgical closure [51]. Thus, impaired wound healing increases morbidity and mortality, prolongs hospitalization and raises healthcare costs. Many of the clinical conditions that increase the risk for impaired wound healing result in tissue hypoxia, or decreased oxygen tension at the wound site. Ultimately, tissue hypoxia impairs immune cell function, reduces granulation tissue production and delays re-epithelialization [53].

One adjuvant strategy that has gained popularity over the last decade is hyperbaric oxygen treatment, or delivery of 100% oxygen at least two times the normal atmospheric pressure at sea level. Inspiration of 100% oxygen increases the partial pressure of oxygen in the plasma and subsequently increases oxygen delivery to vital to healing wounds [54]. Specifically, skin wounds often involve damage to the microvasculature, as well as vasoconstriction, which impair normal oxygen exchange in capillary beds. Thus, a higher partial pressure of oxygen is necessary to force oxygen into the wound site [51]. Application of hyperbaric oxygen creates a large oxygenation gradient between the plasma and the wound site, facilitating oxygen delivery and supporting proper wound healing.

When used as an adjuvant to conventional treatment in non-healing wounds from diabetic subjects, hyperbaric oxygen leads improves wound healing by reducing wound size [55, 56] and accelerating the rate of healing, ultimately reducing the incidence of amputation [57, 58]. In addition to replenishing oxygen, hyperbaric oxygen treatment increases perfusion, prevents ischemia-reperfusion injury and modulates both keratinocyte and immune cell function. Models of epidermal reconstruction illustrate that hyperbaric oxygen thickens both the viable and cornified layers of the skin by enhancing proliferation and differentiation at the cellular level [59]. With exposure to hyperbaric oxygen, keratinocytes show increased expression of the early proliferation marker, p63 and differentiation markers, cytokeratins 1, 10, 11 [59]. Moreover, hyperbaric oxygen also facilitates formation of the dermal scaffold by inducing type collagen IV formation [59]. Collectively, hyperbaric oxygen treatment facilitates several aspects of tissue repair by modulating the physical and chemical components of keratinocyte barrier function.

Conclusion

The formation of the physical and immunologic barriers by the skin depends greatly the coordinated activity of keratinocytes, immune cells, commensal bacteria and various systemic factors. Together, these components help limit the loss of heat, moisture, and vital constituents, while inhibiting invasion of pathogens and exogenous toxins. Following injury, proper wound closure and remodeling depends on carefully orchestrated events that will promote re-epithelialization, re-vascularization and enhanced immunogenicity. Today, our understanding of skin pathologies, including rosacea and psoriasis, have allowed us to begin modulating the interactions between keratinocytes, immune cells, and commensal bacteria in order to improve clinical wound outcomes.

References

[1]	Lebre, M. C., van der Aar, A. M., van Baarsen, L., van Capel, T. M., Schuitemaker, J. H., Kapsenberg, M. L., de Jong, E. C. (2007) Human keratinocytes express functional Toll-like receptor 3, 4, 5, and 9. *J.Invest.Dermatol.* 127, 331-341.

[2]	Baker, B. S., Ovigne, J. M., Powles, A. V., Corcoran, S., Fry, L. (2003) Normal keratinocytes express Toll-like receptors (TLRs) 1, 2 and 5: modulation of TLR expression in chronic plaque psoriasis. *Br.J.Dermatol.* 148, 670-679.

[3]	Kawai, K., Shimura, H., Minagawa, M., Ito, A., Tomiyama, K., Ito, M. (2002) Expression of functional Toll-like receptor 2 on human epidermal keratinocytes. *J.Dermatol.Sci.* 30, 185-194.

[4]	Mempel, M., Voelcker, V., Kollisch, G., Plank, C., Rad, R., Gerhard, M., Schnopp, C., Fraunberger, P., Walli, A. K., Ring, J., Abeck, D., Ollert, M. (2003) Toll-like receptor expression in human keratinocytes:

nuclear factor kappaB controlled gene activation by Staphylococcus aureus is toll-like receptor 2 but not toll-like receptor 4 or platelet activating factor receptor dependent. *J.Invest.Dermatol.* 121, 1389-1396.

[5]   Pivarcsi, A., Bodai, L., Rethi, B., Kenderessy-Szabo, A., Koreck, A., Szell, M., Beer, Z., Bata-Csorgoo, Z., Magocsi, M., Rajnavolgyi, E., Dobozy, A., Kemeny, L. (2003) Expression and function of Toll-like receptors 2 and 4 in human keratinocytes. *Int.Immunol.* 15, 721-730.

[6]   Song, P. I., Park, Y. M., Abraham, T., Harten, B., Zivony, A., Neparidze, N., Armstrong, C. A., Ansel, J. C. (2002) Human keratinocytes express functional CD14 and toll-like receptor 4. *J.Invest.Dermatol.* 119, 424-432.

[7]   Yuki, T., Yoshida, H., Akazawa, Y., Komiya, A., Sugiyama, Y., Inoue, S. (2011) Activation of TLR2 enhances tight junction barrier in epidermal keratinocytes. *J.Immunol.* 187, 3230-3237.

[8]   Lai, Y., Di Nardo, A., Nakatsuji, T., Leichtle, A., Yang, Y., Cogen, A. L., Wu, Z. R., Hooper, L. V., Schmidt, R. R., von Aulock, S., Radek, K. A., Huang, C. M., Ryan, A. F., Gallo, R. L. (2009) Commensal bacteria regulate Toll-like receptor 3-dependent inflammation after skin injury. *Nat.Med.* 15, 1377-1382.

[9]   Yuki, T., Haratake, A., Koishikawa, H., Morita, K., Miyachi, Y., Inoue, S. (2007) Tight junction proteins in keratinocytes: localization and contribution to barrier function. *Exp.Dermatol.* 16, 324-330.

[10]   Helfrich, I., Schmitz, A., Zigrino, P., Michels, C., Haase, I., le Bivic, A., Leitges, M., Niessen, C. M. (2007) Role of aPKC isoforms and their binding partners Par3 and Par6 in epidermal barrier formation. *J.Invest.Dermatol.* 127, 782-791.

[11]   Suzuki, A., Ishiyama, C., Hashiba, K., Shimizu, M., Ebnet, K., Ohno, S. (2002) aPKC kinase activity is required for the asymmetric differentiation of the premature junctional complex during epithelial cell polarization. *J.Cell.Sci.* 115, 3565-3573.

[12]   Pietrocola, G., Arciola, C. R., Rindi, S., Di Poto, A., Missineo, A., Montanaro, L., Speziale, P. (2011) Toll-like receptors (TLRs) in innate immune defense against Staphylococcus aureus. *Int.J.Artif.Organs.* 34, 799-810.

[13]   Kono, H., Rock, K. L. (2008) How dying cells alert the immune system to danger. *Nat.Rev.Immunol.* 8, 279-289.

[14]   Brown, D. L., Kao, W. W., Greenhalgh, D. G. (1997) Apoptosis down-regulates inflammation under the advancing epithelial wound edge:

delayed patterns in diabetes and improvement with topical growth factors. *Surgery.* 121, 372-380.

[15] Kolls, J. K., McCray, P. B.,Jr, Chan, Y. R. (2008) Cytokine-mediated regulation of antimicrobial proteins. *Nat.Rev.Immunol.* 8, 829-835.

[16] Wolk, K., Kunz, S., Witte, E., Friedrich, M., Asadullah, K., Sabat, R. (2004) IL-22 increases the innate immunity of tissues. *Immunity.* 21, 241-254.

[17] Liang, S. C., Tan, X. Y., Luxenberg, D. P., Karim, R., Dunussi-Joannopoulos, K., Collins, M., Fouser, L. A. (2006) Interleukin (IL)-22 and IL-17 are coexpressed by Th17 cells and cooperatively enhance expression of antimicrobial peptides. *J.Exp.Med.* 203, 2271-2279.

[18] Wolk, K., Witte, E., Wallace, E., Docke, W. D., Kunz, S., Asadullah, K., Volk, H. D., Sterry, W., Sabat, R. (2006) IL-22 regulates the expression of genes responsible for antimicrobial defense, cellular differentiation, and mobility in keratinocytes: a potential role in psoriasis. *Eur.J.Immunol.* 36, 1309-1323.

[19] Boniface, K., Bernard, F. X., Garcia, M., Gurney, A. L., Lecron, J. C., Morel, F. (2005) IL-22 inhibits epidermal differentiation and induces proinflammatory gene expression and migration of human keratinocytes. *J.Immunol.* 174, 3695-3702.

[20] Rendon, J. L., Choudhry, M. A. (2012) Th17 cells: critical mediators of host responses to burn injury and sepsis. *J.Leukoc.Biol.* 92, 529-38.

[21] Rendon, J. L., Li, X., Akhtar, S., Choudhry, M. A. (2012) IL-22 Modulates Gut Epithelial and Immune Barrier Functions Following Acute Alcohol Exposure and Burn Injury. *Shock.*

[22] Cash, H. L., Whitham, C. V., Behrendt, C. L., Hooper, L. V. (2006) Symbiotic bacteria direct expression of an intestinal bactericidal lectin. *Science.* 313, 1126-1130.

[23] Zheng, Y., Valdez, P. A., Danilenko, D. M., Hu, Y., Sa, S. M., Gong, Q., Abbas, A. R., Modrusan, Z., Ghilardi, N., de Sauvage, F. J., Ouyang, W. (2008) Interleukin-22 mediates early host defense against attaching and effacing bacterial pathogens. *Nat.Med.* 14, 282-289.

[24] Lai, Y., Li, D., Li, C., Muehleisen, B., Radek, K. A., Park, H. J., Jiang, Z., Li, Z., Lei, H., Quan, Y., Zhang, T., Wu, Y., Kotol, P., Morizane, S., Hata, T. R., Iwatsuki, K., Tang, C., Gallo, R. L. (2012) The antimicrobial protein REG3A regulates keratinocyte proliferation and differentiation after skin injury. *Immunity.* 37, 74-84.

[25] Patel, G. K., Wilson, C. H., Harding, K. G., Finlay, A. Y., Bowden, P. E. (2006) Numerous keratinocyte subtypes involved in wound re-epithelialization. *J.Invest.Dermatol.* 126, 497-502.

[26] Epstein, B., Epstein, J. H., Fukuyama, K. (1983) Autoradiographic study of colchicine inhibition of DNA synthesis and cell migration in hairless mouse epidermis in vivo. *Cell Tissue Kinet.* 16, 313-319.

[27] Levetan, C. S., Upham, L. V., Deng, S., Laury-Kleintop, L., Kery, V., Nolan, R., Quinlan, J., Torres, C., El-Hajj, R. J. (2008) Discovery of a human peptide sequence signaling islet neogenesis. *Endocr.Pract.* 14, 1075-1083.

[28] Kobayashi, S., Akiyama, T., Nata, K., Abe, M., Tajima, M., Shervani, N. J., Unno, M., Matsuno, S., Sasaki, H., Takasawa, S., Okamoto, H. (2000) Identification of a receptor for reg (regenerating gene) protein, a pancreatic beta-cell regeneration factor. *J.Biol.Chem.* 275, 10723-10726.

[29] Aberg, K. M., Radek, K. A., Choi, E. H., Kim, D. K., Demerjian, M., Hupe, M., Kerbleski, J., Gallo, R. L., Ganz, T., Mauro, T., Feingold, K. R., Elias, P. M. (2007) Psychological stress downregulates epidermal antimicrobial peptide expression and increases severity of cutaneous infections in mice. *J.Clin.Invest.* 117, 3339-3349.

[30] Curtis, B. J., Radek, K. A. (2012) Cholinergic regulation of keratinocyte innate immunity and permeability barrier integrity: new perspectives in epidermal immunity and disease. *J.Invest.Dermatol.* 132, 28-42.

[31] Radek, K. A., Elias, P. M., Taupenot, L., Mahata, S. K., O'Connor, D. T., Gallo, R. L. (2010) Neuroendocrine nicotinic receptor activation increases susceptibility to bacterial infections by suppressing antimicrobial peptide production. *Cell.Host Microbe.* 7, 277-289.

[32] Radek, K. A. (2010) Antimicrobial anxiety: the impact of stress on antimicrobial immunity. *J.Leukoc.Biol.* 88, 263-277.

[33] Valera, S., Ballivet, M., Bertrand, D. (1992) Progesterone modulates a neuronal nicotinic acetylcholine receptor. *Proc.Natl.Acad.Sci.U.S.A.* 89, 9949-9953.

[34] Reinheimer, T., Munch, M., Bittinger, F., Racke, K., Kirkpatrick, C. J., Wessler, I. (1998) Glucocorticoids mediate reduction of epithelial acetylcholine content in the airways of rats and humans. *Eur.J.Pharmacol.* 349, 277-284.

[35] Yamasaki, K., Gallo, R. L. (2009) The molecular pathology of rosacea. *J.Dermatol.Sci.* 55, 77-81.

[36] Shibata, M., Katsuyama, M., Onodera, T., Ehama, R., Hosoi, J., Tagami, H. (2009) Glucocorticoids enhance Toll-like receptor 2 expression in

human keratinocytes stimulated with Propionibacterium acnes or proinflammatory cytokines. *J.Invest.Dermatol.* 129, 375-382.

[37] Yamasaki, K., Kanada, K., Macleod, D. T., Borkowski, A. W., Morizane, S., Nakatsuji, T., Cogen, A. L., Gallo, R. L. (2011) TLR2 expression is increased in rosacea and stimulates enhanced serine protease production by keratinocytes. *J.Invest.Dermatol.* 131, 688-697.

[38] Yamasaki, K., Di Nardo, A., Bardan, A., Murakami, M., Ohtake, T., Coda, A., Dorschner, R. A., Bonnart, C., Descargues, P., Hovnanian, A., Morhenn, V. B., Gallo, R. L. (2007) Increased serine protease activity and cathelicidin promotes skin inflammation in rosacea. *Nat.Med.* 13, 975-980.

[39] Sa, S. M., Valdez, P. A., Wu, J., Jung, K., Zhong, F., Hall, L., Kasman, I., Winer, J., Modrusan, Z., Danilenko, D. M., Ouyang, W. (2007) The effects of IL-20 subfamily cytokines on reconstituted human epidermis suggest potential roles in cutaneous innate defense and pathogenic adaptive immunity in psoriasis. *J.Immunol.* 178, 2229-2240.

[40] Christophers, E., Henseler, T. (1987) Contrasting disease patterns in psoriasis and atopic dermatitis. *Arch.Dermatol.Res.* 279 Suppl, S48-51.

[41] Aly, R., Maibach, H. E., Mandel, A. (1976) Bacterial flora in psoriasis. *Br.J.Dermatol.* 95, 603-606.

[42] Singh, G., Rao, D. J. (1978) Bacteriology of psoriatic plaques. *Dermatologica.* 157, 21-27.

[43] Noah, P. W. (1990) The role of microorganisms in psoriasis. *Semin.Dermatol.* 9, 269-276.

[44] Leung, D. Y., Harbeck, R., Bina, P., Reiser, R. F., Yang, E., Norris, D. A., Hanifin, J. M., Sampson, H. A. (1993) Presence of IgE antibodies to staphylococcal exotoxins on the skin of patients with atopic dermatitis. Evidence for a new group of allergens. *J.Clin.Invest.* 92, 1374-1380.

[45] Begon, E., Michel, L., Flageul, B., Beaudoin, I., Jean-Louis, F., Bachelez, H., Dubertret, L., Musette, P. (2007) Expression, subcellular localization and cytokinic modulation of Toll-like receptors (TLRs) in normal human keratinocytes: TLR2 up-regulation in psoriatic skin. *Eur.J.Dermatol.* 17, 497-506.

[46] Morizane, S., Yamasaki, K., Muhleisen, B., Kotol, P. F., Murakami, M., Aoyama, Y., Iwatsuki, K., Hata, T., Gallo, R. L. (2012) Cathelicidin antimicrobial peptide LL-37 in psoriasis enables keratinocyte reactivity against TLR9 ligands. *J.Invest.Dermatol.* 132, 135-143.

[47] Madsen, P., Rasmussen, H. H., Leffers, H., Honore, B., Dejgaard, K., Olsen, E., Kiil, J., Walbum, E., Andersen, A. H., Basse, B. (1991)

Molecular cloning, occurrence, and expression of a novel partially secreted protein "psoriasin" that is highly up-regulated in psoriatic skin. *J.Invest.Dermatol.* 97, 701-712.

[48] Brubaker, A. L., Palmer, J. L., Kovacs, E. J. (2011) Age-related Dysregulation of Inflammation and Innate Immunity: Lessons Learned from Rodent Models. *Aging Dis.* 2, 346-360.

[49] Brubaker, A. L., Schneider, D. F., Kovacs, E. J. (2011) Neutrophils and natural killer T cells as negative regulators of wound healing. *Expert Rev.Dermatol.* 6, 5-8.

[50] Mahbub, S., Brubaker, A. L., Kovacs, E. J. (2011) Aging of the Innate Immune System: An Update. *Curr.Immunol.Rev.* 7, 104-115.

[51] Zamboni, W. A., Browder, L. K., Martinez, J. (2003) Hyperbaric oxygen and wound healing. *Clin.Plast.Surg.* 30, 67-75.

[52] Deitch, E. A., Bridges, R. M., Dobke, M., McDonald, J. C. (1987) Burn wound sepsis may be promoted by a failure of local antibacterial host defenses. *Ann.Surg.* 206, 340-348.

[53] Zhao, L. L., Davidson, J. D., Wee, S. C., Roth, S. I., Mustoe, T. A. (1994) Effect of hyperbaric oxygen and growth factors on rabbit ear ischemic ulcers. *Arch.Surg.* 129, 1043-1049.

[54] Rollins, M. D., Gibson, J. J., Hunt, T. K., Hopf, H. W. (2006) Wound oxygen levels during hyperbaric oxygen treatment in healing wounds. *Undersea Hyperb.Med.* 33, 17-25.

[55] Hammarlund, C., Sundberg, T. (1994) Hyperbaric oxygen reduced size of chronic leg ulcers: a randomized double-blind study. *Plast.Reconstr.Surg.* 93, 829-33; discussion 834.

[56] Zamboni, W. A., Wong, H. P., Stephenson, L. L., Pfeifer, M. A. (1997) Evaluation of hyperbaric oxygen for diabetic wounds: a prospective study. *Undersea Hyperb.Med.* 24, 175-179.

[57] Kalani, M., Jorneskog, G., Naderi, N., Lind, F., Brismar, K. (2002) Hyperbaric oxygen (HBO) therapy in treatment of diabetic foot ulcers. Long-term follow-up. *J.Diabetes Complications.* 16, 153-158.

[58] Faglia, E., Favales, F., Aldeghi, A., Calia, P., Quarantiello, A., Oriani, G., Michael, M., Campagnoli, P., Morabito, A. (1996) Adjunctive systemic hyperbaric oxygen therapy in treatment of severe prevalently ischemic diabetic foot ulcer. A randomized study. *Diabetes Care.* 19, 1338-1343.

[59] Kairuz, E., Upton, Z., Dawson, R. A., Malda, J. (2007) Hyperbaric oxygen stimulates epidermal reconstruction in human skin equivalents. *Wound Repair Regen.* 15, 266-274.

In: Keratinocytes
Editor: Elia Ranzato

ISBN: 978-1-62618-798-6
© 2013 Nova Science Publishers, Inc.

Chapter 7

Keratinocytes and Anti Microbial Peptides: Influence of Extremely Low Frequency Electromagnetic Field (ELF-EMF)

Giovina Vianale[*1], *Matteo Auriemma*[1], *Paolo Amerio*[2] *and Marcella Reale*[1]

[1]Department of Experimental and Clinical Sciences,
[2]Dermatology Clinic,University "G. d'Annunzio",
Chieti-Pescara, Chieti, Italy

Abstract

Keratinocytes are the predominant cells in the epidermis and are involved in the immune response through the secretion of antimicrobial peptides (AMPs). AMPs represent an important evolutionarily conserved innate host-defense mechanism in nearly all organisms. Among the others, human epithelial cells produce mainly β-defensins and chatelicidine.

The functions of defensins in immunomodulation, antimicrobial killing and wound repair have been widely investigated in both human and animal models. Epidemiologic and experimental research on the

---

[*] Email: gvianale@unich.it.

potential effects of extremely low frequency electromagnetic fields (ELF-EMF) on human cells have been performed for a long time. However, there exists a debate on the ability of EMF to influence both inflammatory processes and repair mechanisms including wound healing on different tissue models.

We have demonstrated previously that ELF-EMF increases human keratinocyte cell line (HaCaT) proliferation over time exposure. In the present study, to evaluate the effect of ELF-EMF exposure on β-defensins and pro and anti-inflammatory cytokines expression, HaCaT cells have been exposed to 1 mT, 50 Hz for different lengths of time. Gene expression profiles were analyzed by polymerase chain reaction (PCR) in exposed and unexposed control cells. Preliminary results show that IL-1-β and IL-10 cytokines and HBD-2 and HBD-3 expression were time-dependent regulated by ELF-EMF exposure, whereas,ELF-EMF seem to stabilize VDR expression to control the level in all time points. The relationship between IL1-β, HBD-2 and HBD-3 gene expression seem to be more correlated in ELF-EMF exposed HaCaT cells when compared to non-exposed ones.Taken together, the results induce speculation that ELF-EMF, affecting the cytokines and defensins expression, could speed up and improve tissue repair.

Abbreviations

AD: Atopic Dermatitis
AMPs: Anti Microbial Peptides
ELF-EMF: extremely low frequency electromagnetic fields
PEMF: pulsed-EMF
PCR: polymerase chain reaction
HNP: human neutrophil peptides
IL: Interleukin
PMNs: polymorphonuclear leukocytes.

Introduction

The dermo-epidermal basement membrane zone is the inner sheath of the cutaneous wall. The epidermis protects us against biological, chemical, mechanical and physical injury. It can also defend itself to a certain degree because it is equipped with an evolutionarily old but effective system of primary defense which is referred to as the innate immune system [1].

During evolution, the immune system has developed multifaceted and elaborate mechanisms to defend the host against infections and cancer. The capability of the immune system to distinguish self from non-self is critical in determining when a response will be induced.

Defense mechanisms that are used by the host immediately after encountering a foreign ligand are referred to as innate immunity. These include physical barriers such as the skin and mucosal epithelium, soluble factors such as complement, antimicrobial peptides, chemokines and cytokines, and cells of the innate immune system including monocytes/macrophages, dendritic cells, and polymorphonuclear leukocytes (PMNs) [2]. Keratinocytes, the predominant cells in the epidermis once thought to be inert, can mount an immune response through secretion of antimicrobial peptides. Human epithelial cells produce, among other molecules, a variety of Anti Microbial Peptides (AMPs) that inactivate and kill invading microorganisms [3]. Antimicrobial peptides, in fact, are an important evolutionarily conserved innate host-defense mechanism in many organisms. [4] Previously AMPs were thought to act only as endogenous antibiotics; it is now clear that AMPs trigger and coordinate multiple components of the innate and adaptive immune system [5]. The most common AMPs produced by the skin are Defensins and Chatelicidine. Six cysteine residues forming three intramolecular disulfide bridges characterize Defensins which are divided into α- and β-; there exists six human α-defensins (four named human neutrophil peptides 1–4 and generated by neutrophils; two termed human defensins-5 and -6, produced and stored in the secretory granules of intestinal Paneth cells [6]. Beside α-defensins there are four human β-defensins named h-BD-1 to h-BD-4. Human-BD-1 is expressed constitutively in epidermal keratinocytes and shows antimicrobial activity against predominantly Gram-negative bacteria [1,7]. Human-BD-2, initially isolated from the desquamated scale of psoriatic skin [8] and involved in cutaneous defense and inflammatory mechanisms, is stored in the lamellar bodies characteristically present in the spinous layer of keratinocytes. [9,10] Human-BD-3, cloned from keratinocytes shows a broad spectrum of antimicrobial activity against Gram-negative and Gram-positive bacteria. Its expression in keratinocytes is induced byPathogen-associated molecular patterns or PAMPs, inflammatory mediators such as TNF-α, IL-1β, INF-γ and is differentiated as reviewed elsewhere [11]. The fourth member of the human β-defensin family, h-BD-4 was discovered by screening the human genome databases [12]; similarly to h-BD-2 and h-BD-3, h-BD-4 production is inducible in keratinocytes by inflammatory stimuli, PAMPs and differentiation [11] (Table 1).

**Table 1. Human β-Defensins tissue localization
and modulation of expression [1,7]**

Name	Tissue	Expression
β-Defensins		
H-BD1	Lung, urinary tract, epithelial cells	Constitutively expressed
H-BD2	Skin and other epithelial cells, lung, gut and genitourinary tract	Inducible by: IL-1α, IL-1β, Gram (+), Gram (−), TNF, C. albicans, LPS, LAM, injury, inflammation
H-BD3	Several epithelial surfaces, saliva and vaginal swabs	Inducible by injury and inflammation
H-BD4	Testicles, uterus, stomach	Inducible by injury and inflammation

Expression of antimicrobial peptides is induced upon its encounter with pathogens, during wound healing, in case of epidermal barrier damage and upon increased proinflammatory cytokine production at sites of inflammation either by keratinocytes or other cell types.

Epidemiologic and experimental research on the potential effects of extremely low frequency electromagnetic fields on human cells has been studied for a long time showing contradictory results [13,14].

Only few data are available on the possible influence of EMF on AMPs production. Szymborska-Kajanek et al. [15], have shown that pulsed-EMF (PEMF) do not influence the concentration of defensins in diabetic patients, while it induces their production in healthy subjects. To date, only few data are available on EMF's ability to modulate AMP production. Nevertheless, in diabetic patients, higher levels of those peptides, as well as CRP (C reactive protein) have been shown when compared to healthy controls, confirming the presence of an inflammatory state.

Anti Microbial Peptides

AMPs are small peptidic molecules responsible for the primitive mechanism of immunity, which is present in virtually all-living organisms [1]. Innate Immunity is responsible for the protection against infections, is genetically predetermined and does not require a prior exposure [4]. All tissues and cells exposed to microbes are supposed to produce AMPs [1]. To date more than 1200 AMPs have been described and characterized; the greater part of them share broad-spectrum activity (e.g. direct killing) against bacteria, fungi and viruses [16].

Originally those small peptidic molecules were described to act only as endogenous antibiotics, those with the name AMPs (Anti Microbial Peptides) [17]. Presently, AMPs represent a class of multifunctional peptides able to fight microbes and to modulate multiple components of innate and adaptive immune system [5].

AMPs activity mainly occur as a result of structural characteristics enabling disruption of the microbial membrane, while leaving human cell membranes intact. Besides direct killing, AMPs, act on host cells stimulating cytokine production, cell migration, proliferation, maturation, extracellular matrix synthesis and modulating immune reactions [1,18].

AMPs Classification

Traditionally AMPs can be divided in several different classes such as defensins, cathelicidins and other proteins, mainly produced in different epithelia and inflammatory cells. However, many other proteins not originally included and therefore identified as retaining antimicrobial activities, may function as AMP-like proteins [1].

Two main classes of AMPs exist. Bacteria and fungi produce the first class of AMPs, called bacteriocins, which are well known antibiotics (polymyxin B, streptogramins, vancomycin). The second group contains gene-encoded, ribosomally synthesized oligopeptides or proteins produced by eukaryotic organisms (onward referred to as AMPs).

AMPs characteristically consist of 20 to 60 amino acid residues and carry no unusual post-translational modifications [18] and can be further subdivided in three highly conserved structural patterns (Figure 1):

(1) linear α-helical peptides, free of cysteine residues (chatelicidin/LL-37)
(2) β-sheet globular structures stabilized by intramolecular disulfide bridges (β defensins);
(3) peptides with unusual bias in certain amino acids such as histidine, glycine, proline or tryptophan [1].

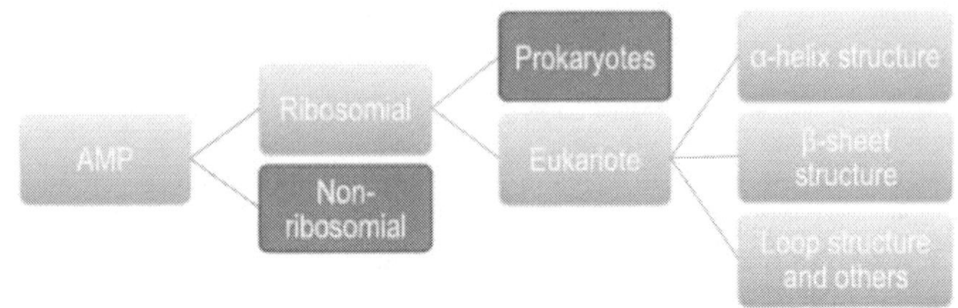

Figure 1. AMPs classification.

AMPs: Something Other Than Antibiotics?

A part from their antibacterial actions, AMPs have different biological effects including: growth stimulus, angiogenesis, protease inhibition, anti-angiogenesis, and chemotaxis [19].

AMPs play an important role in the immune system of mammals, including humans.

As mentioned above, these peptides are ubiquitous in the body exposed to microbes, especially skin and mucosae.

In addition, inflammatory cells as neutrophils, eosinophils and lymphocytes produce large amounts of AMPs. AMPs production may be constitutive, or frequently, induced by inflammation, infections or injury. The different environmental conditions are then responsible for their different production [1].

Cathelicidins, for example, have been shown to drive the repair process in wound healing inducing keratinocyte proliferation and angiogenesis; in addition, cathelicidin retains its antimicrobial function thus assuming a bivalent role in wound repair [4]. AMPs have also been shown to be less inducible in skin from Atopic Dermatitis (AD) patients when compared to normal control and psoriatic skin as a result of the Th2 cytokine milieu, typically present in those patients.

Unbalanced production of antimicrobial peptides has been widely described in different human conditions as in skin inflammatory disorders (psoriasis, atopic dermatitis, rosacea, seborrheic dermatitis), bowel diseases, respiratory and genitourinary tract diseases, as well as in different tumors and in the case of atherosclerosis (Figure 2) [7].

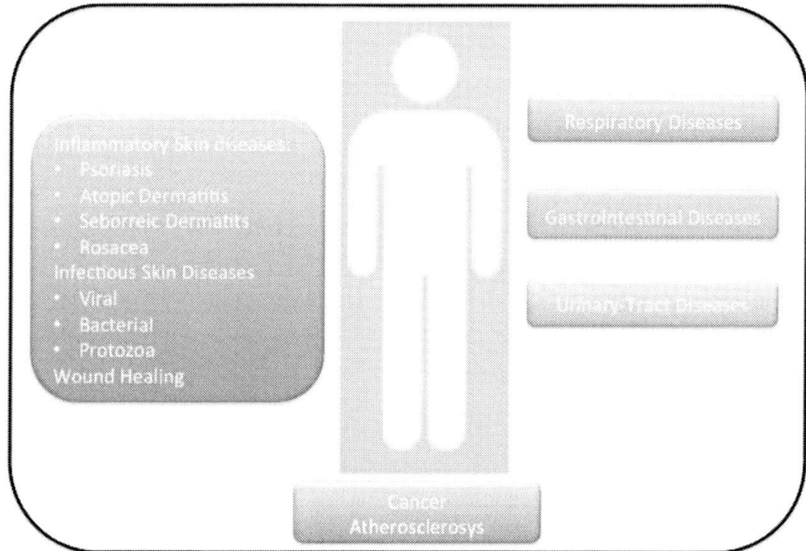

Figure 2. AMPs and Human diseases.

HaCat and EMF

Proliferation

Proliferation and differentiation of keratinocytes are central processes in tissue regeneration after injury. The processes represent the central and final event in tissue regeneration leading to the formation of a massive bulk of cells, necessary to cover the wound area.

Several studies have shown that an electromagnetic field (EMF) can influence both inflammatory processes and repair mechanisms including wound healing on different tissue models [20, 21, 22]. As we previously demonstrated, the electromagnetic field influences the proliferation of keratinocytes (HaCaT) *in vitro* increasing the proliferative activity and reducing the production and the expression of pro-inflammatory molecules such as chemokines and decreasing levels of the transcription factor NF-kB p65 [23].

Immunomodulation

Keratinocytes can also mount an immune response through the secretion of inflammatory cytokines. Keratinocytes constitutively release very low levels of cytokines; however, upon injury or stimulation with exogenous factors such as lipopolysaccharides (LPS), silica, and UV radiation, keratinocytes secrete high levels of IL-1, IL-6, IL-8, IL-10, and TNF-α. These cytokines can induce differentiation and growth of keratinocytes and other resident or migrating cells in the epidermis, dermis, and vessels [24]. Furthermore, they are important mediators of both local and systemic inflammatory and immune responses.

In addition to cytokines, keratinocytes secrete other factors such as neuropeptides, eicosanoids, and reactive oxygen species [25, 26]. These mediators have potent inflammatory and immunomodulatory properties and play an important role in the pathogenesis of cutaneous inflammatory and infectious diseases as well as in aging [27, 28, 29].

HaCat and AMP

Production

Normal human skin receives constant insults from the environment (UVR), as well as mechanical (wounds) and microbial (viruses, bacteria, fungi, protozoa) insults. Those stimuli generate a series of different reactions, among which inflammatory cell recruitment, cytokine and other peptides secretion and proliferation (Figure 3).

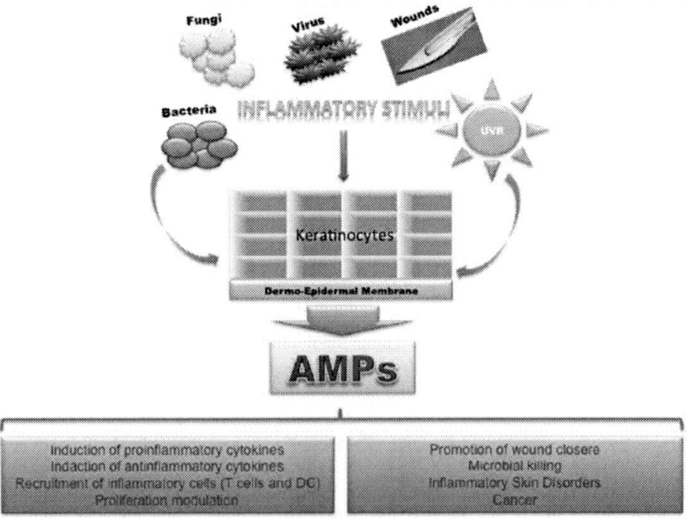

Figure 3. AMPs production.

Keratinocytes produce a wide range of AMPs under different conditions. Increased expression of those peptides have been reported in psoriasis, atopic dermatitis, wounds, infections and other inflammatory skin conditions. For example hBD-1 and hBD-2 mRNA expressions are up-regulated after UVB irradiation, TNF-α and LPS in HaCaT cells [30].

Immunomodulation

AMPs have been widely screened for their immunomodulatory properties. Among which β-defensins have been found to promote adaptive immune response by recruiting dendritic and T cells via CCR6 [31], while hBDs and LL-37 have been reported to enhance the expression of several cytokines in keratinocytes such as IL-6, IL-10, and IP-10. Moreover, LL-37 enhances the expression of monocyte chemoattractant protein-1, macrophage inflammatory protein-3a, RANTES [32] and IL-18 [33] in keratinocytes. Furthermore, human neutrophil peptides (HNP)-1 and -3 stimulate the production of IL-1, IL-4, IL-6, TNF-α, and IFN-γ in monocytes [34].

The immunomodulatory and antibacterial properties of AMPs may be confirmed by the observation that, although UV irradiation has an immunosuppressive action on the skin, it enhances AMPs' production in keratinocytes, thus reducing the risk of bacterial colonization in sun-exposed skin [35].

Human Skin Diseases and AMPs

Different human conditions have been related to an unbalanced AMPs production, of them skin represent one of the more affected system.

Psoriasis patients express higher levels of LL-37 and hBD-2 [36] thus enhancing inflammation via type I interferon production. Patients suffering of rosacea have increased LL-37 levels on skin as well as anomalous proteolytic forms of this peptide [37]. An excessive AMPs secretion (LL-37) from apocrine glands of the hair follicle (HF) infundibulum may trigger HF occlusion and hence the onset of hidradenitis suppurativa [38]. Although there exist contradictory results [39], hBD2 has been described to be over-expressed in Atopic Dermatitis patients and its levels correlate with disease activity [40].

Different mechanisms have been proposed to correlate AMPs expression to the promotion of human skin diseases. Of them, the binding of chatelicidin/LL-37 to self DNA molecules which drives pDC (plasmocytoid Dendritic Cells) to respond mounting an inflammatory response [41], is the better understood mechanism.

In this work we report preliminary data on the possible role of ELF-EMF on defensins production.

Material and Methods

Electromagnetic Field Exposure System

As previously reported [42], all experiments were performed using a sinusoidal 50 Hz electromagnetic field at a flux density of 1 mT (rms) produced by an electromagnetic generator (Hewlett-Packard mod. 33120 with stability higher than 1% both in the frequency and amplitude). A current flow of 1.20 A (I_{eff}) passed through a 160 turn solenoid coil of length l = 0.22 m and radius a = 0.06 m.

The copper wire thickness was 1,25 x 10^{-3} m. The generator was connected to a power amplifier (Denon POA 2800). Cells were located in the central part of the solenoid, which presented the highest degree of field homogeneity (98%). Magnetic field strength and distribution within the solenoid were measured with a Hall-effect probe connected to a Gaussmeter (MG-3D, Walker Scientific Inc., Worcester, MA, USA). Geomagnetic field intensity inside the incubator was about 42 μT; in the same area, the environmental magnetic noise due to power line (50 Hz) was about 7 μT (rms).

The experimental setup for exposure to magnetic field is shown in Figure 4.

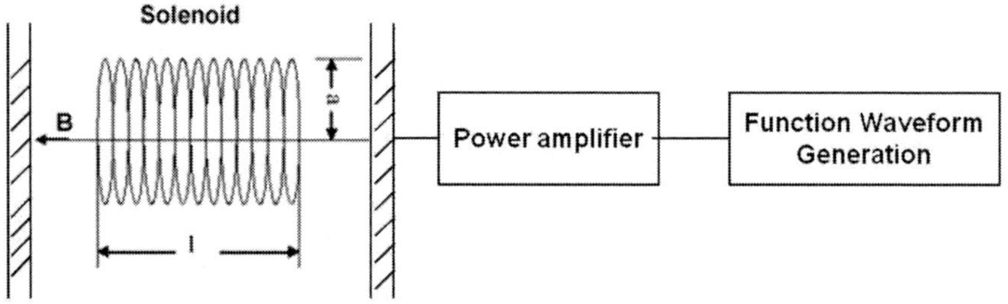

Figure 4. Schematic representation of a typical exposure system. The solenoid was placed inside the cell incubator. The diameter of the solenoid was a = 0.06 m, and its length was l = 0.22 m. The number of solenoid turns was 160, and the diameter of the copper wire was 1.25 · 10-3 m. B indicates the direction of the magnetic field.

Cell Culture and Conditions

HaCaT, immortalized human keratinocytes, were grown in complete medium (DMEM supplemented with 10% heat-inactivated foetal calf serum, 100 U mL^{-1} penicillin and 100 µg mL^{-1} streptomycin). Cultured cells came from the same batch and incubated at 37°C in a humidified atmosphere of 5% CO_2 for both control and exposed cultures. HaCaT cells were exposed to 50 Hz electromagnetic field at a flux density of 1 mT (rms) produced by an electromagnetic generator. Cell flasks were located on a grid to allow the correct air circulation in the central part of the solenoid that presented the highest degree of field homogeneity (98%). Non-exposed cells were placed in a different incubator far from the solenoid, where only environmental magnetic field can be detected. The temperature inside the incubator was monitored by the built-in thermometer (T = 37 ± 0,2 °C) and by an additional digital thermometer (HD 9216, Delta OHM, Italy) attached to plates during exposure to quantify local temperature variations (ΔT = 0,13 °C). The low-level Joule heating was efficiently dissipated inside the incubator by the fan system.

Cell viability was not influenced by field exposure. More than 98% of cells were viable, as determined by Trypan blue dye exclusion at the beginning of the culture, and more than 90% were viable before cells were collected.

RNA Extraction, Reverse Transcription–Polymerase Chain Reaction (RT-PCR)

Total cellular RNA was extracted from cellsusing Trizol (Invitrogen, Carlsbad, CA, USA) according to the manufacturer's instructions. RNA samples were examined for concentration, purity and integrityusing NanoDrop Technologies (Thermo Scientific, USA). For cDNA synthesis, to avoid genomic DNA contamination during RT-PCR, 1 µg total RNA was retrotranscribed with the QuantiTect Reverse Transcription Kit (QIAGEN GmbH, Hilden, Germany) according to the manufacturer's recommendations; the kit achieves complete digestion of genomic DNA by brief sample incubation with a specific Wipeout buffer at 42 °C before retrotranscription. cDNA was then stored at -20 °C until use. GenBank accession numbers were used to design primers for the genes of interest and the housekeeping genes. Primer details, and the amplicon length of each gene examined are reported in Table 2. Primers were designed using AlleleID (PREMIER Biosoft

International, Palo Alto, CA, USA) and IDT SciTools (Tema Ricerca, Bologna, Italy). Special care was devoted to primer length, annealing temperature, base composition **and** 3'-end stability. Products were subsequently run on 2% agarose gel to check for size specificity. The PCR amplification program was as follows: initial denaturation at 95°C for 5 min, followed by 30 cycles of denaturation at 95°C for 30 s, annealing at 60°C for 30 s and 72°C for 30 s, extension and a final step at 72°C for 10 min. All the RT-PCR was performed three times.

Each RT-PCR product was run onto 2% agarose gel stained with ethidium bromide, the gels were then scanned and acquired to perform densitometric analysis and determine the fold induction with Fluor Chem FC2 (M.Medical, Mi, Italy).

Gene expression was evaluated for IL-1β, IL-10, HBD-2, HBD-3 and VDR, and normalized with 18S. PCR amplification products of 18S reflect the amount of cDNA input. Normalized expression ratios are depicted graphically.

Table 2. Primer sequences of the explored genes
(Forward and Reverse sequences)

Gene	FW primer	RW primer
IL-1b	TGAGGATGACTTGTTCTTTGAAG	GTGGTGGTCGGAGATTCG
IL-10	GAGAACCAAGACCAGACATC	TCACTCATGGCTTTGTAGATGC
HBD-2	TCAGCCATGAGGGTCTTGTA	GGATCGCCTATACCACCAAA
HBD3	CCTGTTTTTGGTGCCTGTTC	CACTCTCGTCATGTTTCAGGG
VDR	CACTATTCACCTGCCCCTTC	CTTCCTCTGCACTTCCTCATC
18S	TTGCCATCACTGCCATTAAG	TTTCCATCCTTTACATCCTTCTG

Results

ELF-EMF has shown to modulate the expression of immunomodulatory and proinflammatory cytokines and peptides in HaCaT cell cultures in experimental conditions. ELF-EMF was shown to upregulate immunomodulatory cytokine expression, as IL10 (a prototypic immunosuppressive cytokine), especially in the first hours of culture, when compared to non-exposed cells.

While IL-10 expression rises, IL-1β expression is sensibly reduced in the exposed cells. Both cytokines reached control levels in the longer exposure timepoints (36h) (Data not shown).

H-BD2

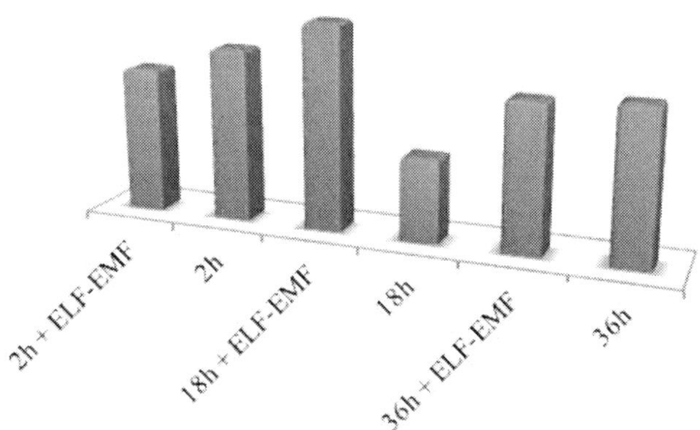

Figure 5. H-BD2 gene expression in HaCaT cells inside and outside ELF-EMF. Gene expression has been 18S/normalized.

H-BD3

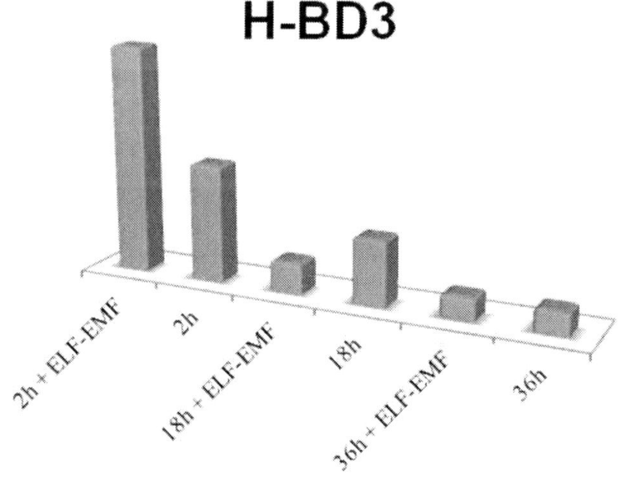

Figure 6. H-BD3 gene expression in HaCaT cells inside and outside ELF-EMF. Gene expression was 18S/normalized.

Interestingly, human beta defensin 2 (HBD-2) (*Figure 5*) and human beta defensin 3 (HBD-3) (*Figure 6*) was induced in ELF-EMF exposed HaCaT cells when compare to non-exposed cells. However this induction was not sustained over all the time points, thus implying that ELF-EMF does not sustain autoinflammatory *stimuli* at least in cultures of HaCaT cells.

Interestingly the induction of h-BD's expression seems to be independent from the VDR pathway, opening a window on a possible new modulation pathway of h-BD expression.

Conclusion

In this work, we sought to elucidate the role of ELF-EMF on the cultured keratinocyte cell line (HaCaT). EMF exposure confirmed induction of a relative immunosuppressed state in those cells, enhancing the expression of IL-10, a prototypic immunosuppressive cytokine, and reducing the expression of IL-1.

Interestingly, besides the immunosuppressive effect, we can hypothesize that ELF-EMF induces the expression of specific peptides that are part of the ancient arm of immunity. Human-DB2 and human-BD3 were shown to be upregulated in exposed HaCaT cells when compared to non-exposed ones. H-BD initiate the inflammatory mechanisms in several conditions: infections, allergic diseases, inflammatory diseases and tissue regeneration. We already showed the ability of ELF-EMF to induce HaCaT cells proliferation while reducing the expression of proinflammatory molecules in a previous paper [23]. Currently, this mechanism is still under investigation. With this work we want to extend the understanding of the effects of electromagnetic fields on cell lines. On the basis of our experiments it is possible to speculate that ELF-EMF improve the proliferative rate by inducing the expression of antinflammatory cytokines and of h-BD2 and h-BD3 while reducing IL-1β in a VDR independent manner. More experiments are necessary to better elucidate those mechanisms and to verify any other pathways responsible for the increased proliferative rate in EMF expose cells.

References

[1]	Wiesner J, Vilcinskas A. Antimicrobial peptides: the ancient arm of the human immune system. *Virulence* 2010; 1: 440-64.

[2]	Yuping Lai and Richard L. Gallo AMPed Up immunity: how antimicrobial peptides have multiple roles in immune defense. *Trends Immunol.* 2009; 30(3): 131–141.

[3] Lehrer IR, Ganz T. Antimicrobial peptides in mammalian and insect host defense. *Curr. Opin. Immunol.* 1999;11(1):23-7.

[4] Izadpanah A, Gallo RL. Antimicrobial peptides. *J. Am. Acad. Dermatol.* 2005; 52:381-90.

[5] Schauber J, Gallo RL. Antimicrobial peptides and the skin immune defense system. *J. Allergy Clin. Immunol.* 2009; 124: R13.

[6] Oppenheim JJ, Biragyn A, Kwak LW, Yang D. Roles of antimicrobial peptides such as defensins in innate and adaptive immunity. *Ann. Rheum. Dis.* 2003; 62:17–21.

[7] Guaní-Guerra E, Santos-Mendoza T, Lugo-Reyes SO, Terán LM. Antimicrobial peptides: general overview and clinical implications in human health and disease. *Clin. Immunol.* 2010; 135(1):1-11.

[8] Harder J, Bartels J. Christophers E. Schroder JM. A peptide antibiotic from human skin. *Nature.* 1997; 387(6636): 861.

[9] Ali RS, Falconer A, Ikram M, Bissett CE, Cerio R, Quinn AG. Expression of the peptide antibiotics human beta defensin-1 and human beta defensin-2 in normal human skin. *J. Invest Dermatol.* 2001; 117(1): 106-11.

[10] Chadebech P, Goidin D, Jacquet C, Viac J, Schmitt D, Staquet MJ. Use of human reconstructed epidermis to analyze the regulation of beta-defensin hBD-1, hBD-2, and hBD-3 expression in response to LPS. *Cell Biol. Toxicol.* 2003; 19(5): 313-24.

[11] Pivarcsi A, Nagy I, Kemeny L. Innate Immunity in the Skin: How Keratinocytes Fight Against Pathogens. *Current Immunology Reviews,* 2005; 1, 29-42.

[12] García JR, Krause A, Schulz S, Rodríguez-Jiménez FJ, Klüver E, Adermann K, Forssmann U, Frimpong-Boateng A, Bals R, Forssmann WG. Human beta-defensin 4: a novel inducible peptide with a specific salt-sensitive spectrum of antimicrobial activity. *FASEB J.* 2001;15(10):1819-21.

[13] Simkó M, Mattsson MO. Extremely low frequency electromagnetic fields as effectors of cellular responses in vitro: possible immune cell activation. *J. Cell Biochem.* 2004; 1;93(1):83-92.

[14] Cook CM, Thomas AW, Prato FS. Human electrophysiological and cognitive effects of exposure to ELF magnetic and ELF modulated RF and microwave fields: a review of recent studies. *Bioelectromagnetics.* 2002;23(2):144-57.

[15] Szymborska-Kajanek A, Strzelczyk JK, Karasek D, Rawwash HA, Biniszkiewicz T, Cieślar G, Hajdrowska B, Sieroń-Stołtny K, Sieroń A,

Wiczkowski A, Grzeszczak W, Strojek K. Impact of Low-frequency Pulsed Magnetic Fields on Defensin and CRP Concentrations in Patients with Painful Diabetic Polyneuropathy and in Healthy Subjects. *Electromagn. Biol. Med.* 2010;29(1-2):19-25.

[16] Kreuter A, Jaouhar M, Skrygan M, Tigges C, Stücker M, Altmeyer P, Gläser R, Gambichler T. Expression of antimicrobial peptides in different subtypes of cutaneous lupus erythematosus. *J. Am. Acad. Dermatol.* 2011;65(1):125-33.

[17] Lai Y, Gallo RL. AMPed up immunity: how antimicrobial peptides have multiple roles in immune defense. *Trends Immunol.* 2009;30(3):131-41.

[18] Kenshi Y, Gallo RL. Antimicrobial peptides in human skin disease. *Eur. J. Dermatol.* 2008; 18(1): 11–21.

[19] Pasupuleti M, Roupe M, Rydengård V, Surewicz K, Surewicz WK, Chalupka A, Malmsten M, Sörensen OE, Schmidtchen A.Antimicrobial activity of human prion protein is mediated by its N-terminal region. *PLoS One.* 2009 Oct 7;4(10):e7358.

[20] Kheifets L, Shimkhada R. Childhood leukemia and EMF: review of the epidemiologic evidence. Bioelectromagnetics 2005; 7:S51–9. Bassett CA. Beneficial effects of electromagnetic fields. *J. Cell Biochem.* 1993; 51:387–93.

[21] Gmitrov J, Ohkubo C, Okano H. Effect of 0.25 T static magnetic field on microcirculation in rabbits. *Bioelectromagnetics* 2002; 23: 224–9.

[22] Tenuzzo B, Chionna A, Panzarini E et al. Biological effects of 6 mT magnetic fields: a comparative study in different cell types. *Bioelectromagnetics* 2006; 27:560–70.

[23] Vianale G, Reale M, Amerio P, Stefanachi M, Di Luzio S, Muraro R. Extremely low frequency electromagnetic field enhances human keratinocyte cell growth and decreases proinflammatory chemokine production. *Br. J. Dermatol.* 2008 Jun;158:1189-96.

[24] Werner S., Grose R. Regulation of Wound Healing by Growth Factors and Cytokines. *Physiol. Rev.* 2003; 83: 835–870.

[25] Rousseau K, Kauser S, Pritchard LE, Warhurst A, Oliver RL, Slominski A, Wei ET, Thody AJ, Tobin DJ, White A. Proopiomelanocortin (POMC), the ACTH/melanocortin precursor, is secreted by human epidermal keratinocytes and melanocytes and stimulates melanogenesis. *FASEB J.* 2007 Jun;21(8):1844-56.

[26] Sjursen W, Brekke OL, Johansen B. Secretory and cytosolic phospholipase A(2)regulate the long-term cytokine-induced eicosanoid production in human keratinocytes. *Cytokine.* 2000;12(8):1189-94.

[27] Glassman SJ. Vitiligo, reactive oxygen species and T-cells. *Clin. Sci. (Lond).* 2011;120(3):99-120.

[28] **Poljšak B, Dahmane R. Free radicals** and extrinsic skin aging. *Dermatol. Res. Pract.* 2012;2012:135206.

[29] Grange PA, Chéreau C, Raingeaud J, Nicco C, Weill B, Dupin N, Batteux F. Production of superoxide anions by keratinocytes initiates P. acnes-induced inflammation of the skin. *PLoS Pathog.* 2009;5(7):e1000527.

[30] Seo SJ, Ahn SW, Hong CK, Ro BI. Expressions of beta-defensins in human keratinocyte cell lines. *J. Dermatol. Sci.* 2001;27(3):183-91.

[31] Yang D, Chertov O, Bykovskaia SN, Chen Q, Buffo MJ, Shogan J, Anderson M, Scho¨rder JM, Wang JM, Howard OMZ, Oppenheim JJ. -defensins: linking innate and adaptive immunity through dendritic and T cell CCR6. *Science* 1999;15, 286:525–8.

[32] Niyonsaba F, Ushio H, Nakano N, Ng W, Sayama K, Hashimoto K, Nagaoka I, Okumura K, Ogawa H. Antimicrobial peptides human beta-defensins stimulate epidermal keratinocyte migration, proliferation and production of proinflammatory cytokines and chemokines. *J. Invest. Dermatol.* 2007 Mar;127(3):594-604.

[33] Niyonsaba F, Ushio H, Nagaoka I, Okumura K, Ogawa H (2005) The human b-defensins (hBD-1, -2, -3, -4) and cathelicidin LL-37 induce interleukin-18 secretion through p38 and ERK MAPK activation in primary human keratinocytes. *J. Immunol.* 175:1776–84.

[34] Chaly YV, Paleolog EM, Kolesnikova TS, Tikhonov II, Petratchenko EV, Voitenok NN (2000) Neutrophil a-defensin human neutrophil peptide modulates cytokine production in human monocytes and adhesion molecule expression in endothelial cells. *Eur. Cytokine Netw.* 11:257–66.

[35] Gläser R, Navid F, Schuller W, Jantschitsch C, Harder J, Schröder JM, Schwarz A, Schwarz T. UV-B radiation induces the expression of antimicrobial peptides in human keratinocytes in vitro and in vivo. *J Allergy Clin. Immunol.* 2009;123(5):1117-23.

[36] Ong PY, Ohtake T, Brandt C, Strickland I, Boguniewicz M, Ganz T, Gallo RL, Leung DY. Endogenous antimicrobial peptides and skin infections in atopic dermatitis. *N. Engl. J. Med.* 2002;10;347(15):1151-60.

[37] Yamasaki K, Di Nardo A, Bardan A, Murakami M, Ohtake T, Coda A, Dorschner RA, Bonnart C, Descargues P, Hovnanian A, Morhenn VB,

Gallo RL. Increased serine protease activity and cathelicidin promotes skin inflammation in rosacea. *Nat. Med.* 2007;13(8):975-80.

[38]	Emelianov VU, Bechara FG, Gläser R, Langan EA, Taungjaruwinai WM, Schröder JM, Meyer KC, Paus R. Immunohistological pointers to a possible role for excessive cathelicidin (LL-37) expression by apocrine sweat glands in the pathogenesis of hidradenitis suppurativa/acne inversa. *Br. J. Dermatol.* 2012;166(5):1023-34.

[39]	Yamasaki K, Gallo RL. Antimicrobial peptides in human skin disease. *Eur. J. Dermatol.* 2008;18(1):11-21.

[40]	Kanda N, Watanabe S. Increased serum human β-defensin-2 levels in atopic dermatitis: relationship to IL-22 and oncostatin M. *Immunobiology.* 2012; 217(4):436-45.

[41]	Lande R, Gregorio J, Facchinetti V, Chatterjee B, Wang YH, Homey B, Cao W, Wang YH, Su B, Nestle FO, Zal T, Mellman I, Schröder JM, Liu YJ, Gilliet M. Plasmacytoid dendritic cells sense self-DNA coupled with antimicrobial peptide. *Nature.* 2007:4;449(7162):564-9.

[42]	Reale M, De Lutiis MA, Patruno A, Speranza L, Felaco M, Grilli A, Macrì MA, Comani S, Conti P, Di Luzio S. Modulation of MCP-1 and iNOS by 50-Hz sinusoidal electromagnetic field. *Nitric Oxide* 2006;15: 50–57.

Index

C

D

N

O